高等院校土建学科双语教材（中英文对照）

◆ 土木工程专业 ◆

BASICS

承重结构

LOADBEARING SYSTEMS

[德] 阿尔弗莱德·梅斯特曼　编著

文捷　译

中国建筑工业出版社

著作权合同登记图字：01－2007－3334号

图书在版编目（CIP）数据

承重结构/（德）梅斯特曼编著；文捷译.—北京：中国建筑工业出版社，2011
高等院校土建学科双语教材（中英文对照）◆土木工程专业◆
ISBN 978－7－112－11596－9

Ⅰ.承… Ⅱ.①梅…②文… Ⅲ.建筑结构－结构载荷－高等学校－教材－汉、英 Ⅳ.TU312

中国版本图书馆CIP数据核字（2009）第210928号

责任编辑：孙　炼
责任设计：董建平
责任校对：兰曼利

高等院校土建学科双语教材（中英文对照）
◆土木工程专业◆
承重结构
［德］阿尔弗莱德·梅斯特曼　编著
文捷　译
*
中国建筑工业出版社出版、发行（北京西郊百万庄）
各地新华书店、建筑书店经销
北京嘉泰利德公司制版
北京建筑工业印刷厂印刷
*
开本：880×1230毫米　1/32　印张：4¾　字数：136千字
2011年5月第一版　2011年5月第一次印刷
定价：**16.00**元
ISBN 978－7－112－11596－9
（20278）

中文部分目录

\\ 前言 6

\\ 荷载和力 86
- \\ 承重结构和静力学 86
- \\ 力 86
- \\ 静力系统 87
- \\ 外力 88
- \\ 内力 94
- \\ 确定尺寸 103

\\ 结构构件 108
- \\ 悬臂梁、简支梁和带悬臂简支梁 108
- \\ 连续梁 110
- \\ 铰接梁 110
- \\ 桁架梁 113
- \\ 格构 114
- \\ 板 116
- \\ 柱 118
- \\ 索 120
- \\ 拱 123
- \\ 刚架 125

\\ 承重结构 128
- \\ 实体结构 128
- \\ 框架结构 131
- \\ 支撑构件 136
- \\ 大厅 139
- \\ 板结构 142
- \\ 基础 144

\\ 结语 147

\\ 附录 148

\\ 确定构件尺寸的原则 148
\\ 参考文献 150
\\ 图片说明 150

CONTENTS

\\Foreword _7

\\Loads and forces _9
\\Loadbearing structures and statics _9
\\Forces _9
\\Statical system _10
\\External forces _11
\\Internal forces _18
\\Dimensioning _28

\\Structural elements _35
\\Cantilever arm, simply supported beam, simply supported beam with cantilever arm _35
\\Continuous beam _37
\\Articulated beam _38
\\Trussed beam _40
\\Lattice _41
\\Slab _44
\\Column _46
\\Cable _49
\\Arch _51
\\Frame _54

\\Loadbearing structures _59
\\Solid construction _59
\\Skeleton construction _62
\\Reinforcement _68
\\Halls _70
\\Plate structures _74
\\Foundations _77

\\In conclusion _81

\\Appendix _82
\\Pre-dimensioning formulae _82
\\Literature _84
\\Picture credits _84

前　言

建造建筑时，我们需要知道其结构性能如何。承重结构可以是设计中的主导因素，也可能是我们平时看不见的辅助结构，但无论如何建筑总是要基于这些承重结构的。它将结构组合在一起，将荷载传递到地面并保持结构的稳定性。对于承重结构的理解，包括结构基本原理和各种承重系统的特性，是将这些原理合理地应用于设计并找到合适的材料及建造方法的基础。

课程的开始通常是很困难的，特别是要消化这么多的新材料，了解复杂的静力学以及承载理论。承重结构基础将建筑学科与土木工程学科联系在一起，并深入浅出地全面阐述了承重结构理论的基本知识。为了使读者更好地理解，作者首先用示例和简单的背景知识解释了建筑内的荷载和力，介绍设计中可能用到的典型承重结构构件及不同类型的承重系统和结构。这里介绍的简明知识使得学生可以完整地认识承重结构并进行创造性的设计。

编者：Bert Bielefeld

FOREWORD

When constructing a building, we need to know how its structural properties function. Loadbearing elements can be the dominant features of the design, or simply an invisible substructure – but a building is always based on its loadbearing structure. It holds the building together, distributes loads into the ground, and guarantees stability. An understanding of loadbearing – its structural principles and the specific qualities of individual loadbearing systems – is fundamental to applying these principles sensibly in the design process and developing a solution that suits the materials and the construction method.

It is often difficult, particularly at the beginning of a course when there is so much new material to assimilate, to work one's way into the complexities of statics and loadbearing theory. *Basics Loadbearing Systems* bridges the fields of architecture and civil engineering and explains the fundamentals of loadbearing structure theory simply, comprehensibly and chronologically. To help general understanding, the author first explains the loads and forces occurring in a building using examples and simple contexts. He introduces typical loadbearing structural elements and shows loadbearing systems and structures for the different building types that planners can use for their designs. The compact knowledge conveyed here makes it possible for students to work with loadbearing structures in an integrated way, and thus be able to design creatively.

Bert Bielefeld, Editor

LOADS AND FORCES

LOADBEARING STRUCTURES AND STATICS

A great deal of philosophizing can be done about how design relates to construction. Very different positions can be taken, but they are always two sides of the same coin. Designing spaces means defining them, by applying theory to structures that will need to be realized. Knowing about structures is therefore one of the fundamentals of architectural theory. It is very rare for the architect him- or herself to vouch for the stability of constructions. But he or she should be in a position to select structural elements correctly at the early design stages and to assess the dimensions needed for them realistically. The next step is usually to develop the loadbearing system with a structural engineer. To be able to work together effectively, fundamental knowledge is needed about loadbearing systems and structures, their advantages and disadvantages and the forces that come into play. These different forces seem complex at first, but they are logically coherent.

It is easiest to explain how they fit together in the order in which they are addressed for a statical calculation. A calculation of this type usually follows these steps:

- Analysing the overall structure and the function of the individual structural elements in it – statical system
- Determining all the forces working on the structural elements – assumed loads
- Calculating the forces affecting a particular structural element and the forces that it transmits to others – calculating the external forces
- Calculating the forces within the structural element itself – investigating internal or static forces
- Determining the stability of the planned structural element
- Proof that the planned structural element can withstand the forces determined

FORCES

$F = m \cdot a$

Force is defined as mass times acceleration.

Newton

The unit used for measuring force is the newton; a newton corresponds roughly to the weight of 100 grammes. In building the newton is complemented by the kilonewton and the meganewton.

Kilonewton: 1 kN = 1000 N, Meganewton: 1 MN = 1,000,000 N

A force is determined by magnitude and direction. Its action is linear, and is expressed by its line of application and the direction of this line. › Fig. 1

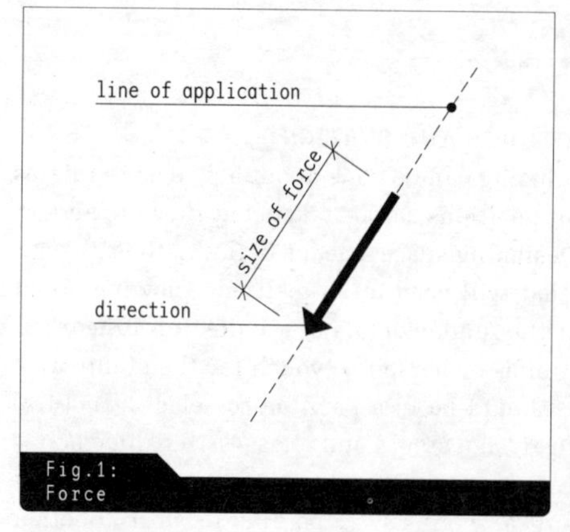

Fig.1: Force

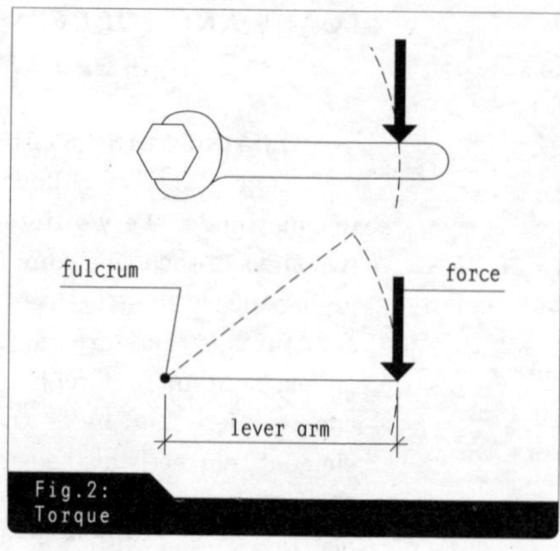

Fig.2: Torque

Moments, torque

Forces can also work in a circle around a point. They are then called torque or moments, and are defined by their size multiplied by the distance from the fulcrum (lever arm).

A simple example of torque is tightening a screw with a screwdriver. This also demonstrates the link between force magnitude and lever arm. The longer the lever arm, the greater the torque. › Fig. 2

Action = reaction

Statics describes the distribution of forces in a system at rest. Buildings or parts of buildings are usually motionless, and all the effective forces balance each other out. This can be summed up in the law "action = reaction". It is used as a starting point in statical calculations, on the basis that the sum of all forces in any one direction and its counter-direction is zero. If the action is known, the reaction can be determined immediately. The chapter External forces, Support forces explains the methods that apply this possible to loadbearing systems.

STATICAL SYSTEM

A structural engineer first establishes the connections within the construction in the statical system. A statical system is an abstract model of the real, complex structure of the component parts. Supporting members are considered as lines even if they have a wide cross section, and their load is treated as a point. Walls are presented as disc structures and their loads are applied in lines. Additional information the statical system gives is how the structural elements are joined together, and how their forces are distributed from one element to another. This is crucial to the calculations. The symbols used in statical systems are explained in the chapter External forces, Support forces › Fig. 8, p. 16 and are used subsequently in the text.

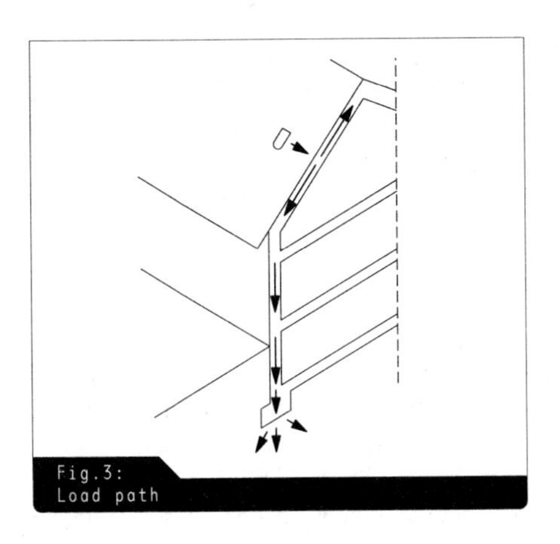

Fig.3: Load path

Positions

The next working stage involves identifying all the structural elements in sequence as positions and numbering them. Here it is also important to establish which structural elements load which others.

Load path

For example, roof tiles are not just supported by the roof structure, but also affect the walls, right down to the foundations. It must be established with absolute precision which structural elements absorb the loads from the upper storeys. › Fig. 3

›✎

EXTERNAL FORCES

If we consider a building element such as a roof beam, we distinguish between two types of force. First, there are the forces exerted on it by the roof structure above it, and those that it transfers to the masonry supporting it. If we do not consider its dead weight, it does not matter in the first

✎

\\Tip:
For good cooperation with structural engineers, it is important that designers be familiar with these specialists' part of the work in a project and understand their working methods and aims. It therefore makes sense to look at their calculations and positional and working plans and compare them with the architect's documents. After the structural engineer has devised the structure with the architect in the design phase, the main thrust of his or her work is to draw up the statics for planning permission and later to draw the plans for constructing the shell. Here the interest is above all in the loadbearing parts of a building. All the non-loadbearing elements, even non-loadbearing walls, for example, are only significant as loads, and may not feature in the plans at all.

place whether this beam is thick or thin, weak or strong, as we are dealing with external forces that do not include the beam itself.

We must distinguish between external forces and the internal forces operating in the beam itself. For example, how great is the bending force in the roof beam exerted by the roof construction it supports? This bending moment is one of the internal forces that will be explained in the corresponding chapter.

Actions

Everything that can affect a structural element is called an action. Actions are usually forces with different causes. Forces that affect structural elements mechanically are also called loads.

Loads

Loads affect structural elements from the outside, and we must distinguish between them and the reaction forces explained in the subsection Support forces. Loads are divided into various categories. We distinguish between point, line and area loads, according to the degree of abstraction of the statical system. › Fig. 4

In addition, we distinguish between constant, variable and extraordinary actions, in relation to the duration of the action.

Permanent loads

Constant action includes, above all, the weight forces of the structural elements, called permanent loads.

Working loads

The working loads include the variable actions wind, snow and ice loads. Working loads have to be planned in at standard levels for the building's intended use. The most important are the vertical working loads that must be worked out for floors. Whether the rooms are for homes, offices, meeting rooms or some other purpose, they must be given an appropriate working load value as an area load. Largely horizontally applied loads also have to be taken into account, such as loads on railings and parapets, braking, acceleration and collision loads for vehicles, dynamic loads for machines, and earthquake loads. The size of these loads is fixed in national standards, which give them in tables. › Appendix, Literature

Assumed loads

After using the statical system to explain how the structure functions, the next step is to determine the actions. All the acting forces must be identified, assigned a value and added together. They are generally related to a metre or square metre of the structural element. Loads acting obliquely are usually divided into a horizontal and a vertical element.

Vertical load
Horizontal load
›

For further calculations we distinguish between vertical loads, horizontal loads and torque.

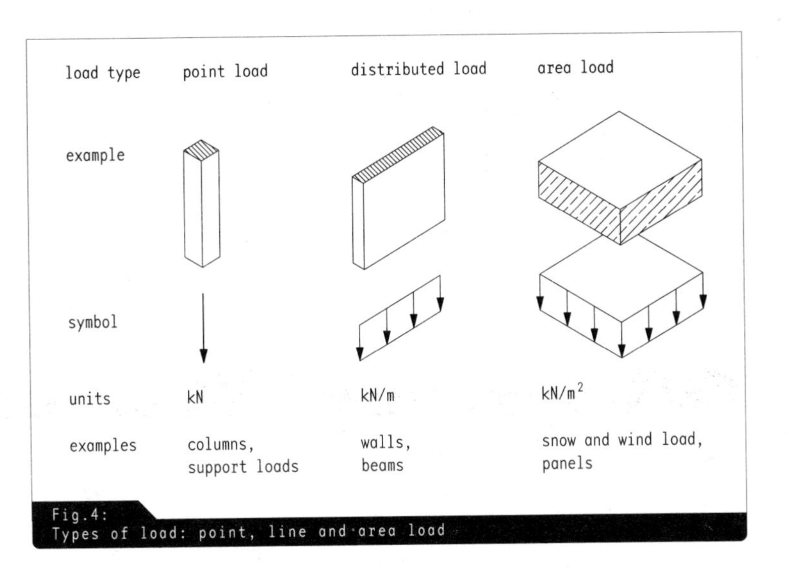

Fig.4:
Types of load: point, line and area load

Load absorption area

Load absorption area describes the particular reference area for loads on a structural element. It is part of an overall surface whose load is being dissipated to a certain structural element. It relates to the nature and span of a structure.

Example: The beams of a timber beam floor are 80 cm apart. Which part of the floor is acting on an individual beam? The load absorption area extends from the middle of the space between the beams on the left-hand side to the middle of the space on the right-hand side, twice 40 cm. So overall it is again 80 cm wide. › Fig. 5 This is a simple example, but determining

\\Important:
Loads acting vertically per square metre in a structural element: dead weight, working loads for floors, stairs, balconies
Acting vertically per square floor plan metre: snow load
Acting at right angles to the area of the structural element: wind load
Generally acting horizontally: loads on parapets and railings, braking and acceleration loads, collision loads from vehicles, earthquake loads

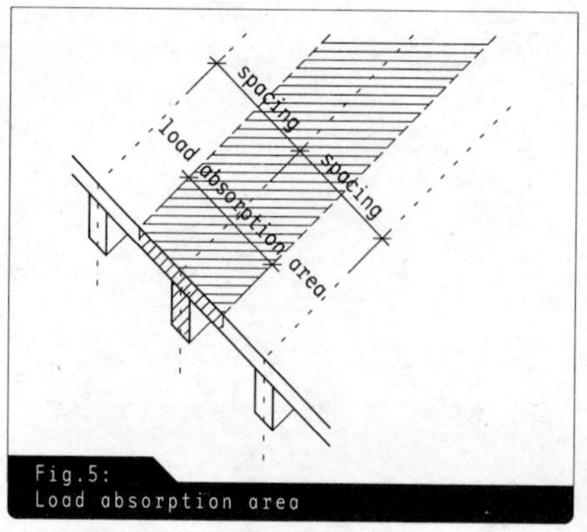

Fig.5:
Load absorption area

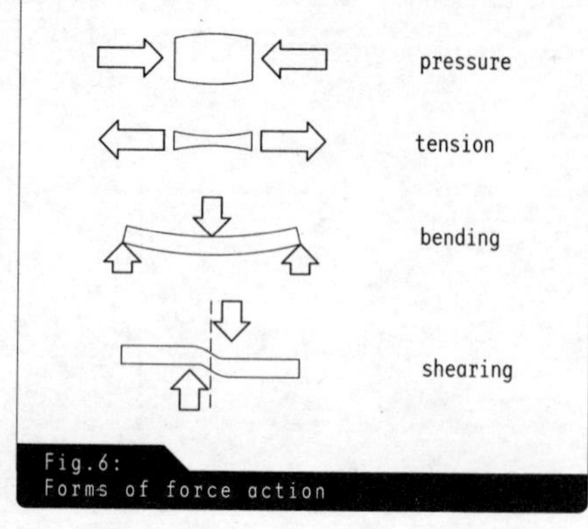

Fig.6:
Forms of force action

the load absorption area can be more complicated according to the particular structural element.

Force action forms

So far we have considered loads and their magnitude, but how a load, or more generally a force, acts on a structural element is also important. Here we distinguish between the following action forms:

- Compression: one stone lies on top of another, exerting pressure on it.
- Tension: tensile load is most clearly explained using the example of a rope, which can absorb only tensile forces.
- Bending: a beam is fixed at both ends and then loaded from above. It sags, i.e. it is subject to a bending load.
- Shearing: this load is explained by the way a pair of household scissors loads paper to cut it. Two forces work on each other slightly offset and transversely to the structural element. This load often acts on connecting devices such as screws. › Fig. 6

Supports

Points of contact between structural elements at which forces are transmitted are called supports. A simple example is a ceiling beam supported on masonry. The beam has its support on the crown of the wall. In building the idea of the support is somewhat broader, and covers many different points of contact between structural elements. For example, when a flagpole is fixed into the ground or a steel beam is connected to a steel

Fig.7:
Supports in steel construction

column, this is also called a support. In terms of structural engineering they differ primarily in the forces that they can dissipate.

It is very easy to look at the different forms of supports in old steel bridges. Large bridge girders are supported on very small points or narrow strips. This means that the girders can deflect without interference from the supports, which are then known as articulated supports. These are used on one side of the bridge, while those on the other side are additionally supported by steel rollers.

Expansion bearings

When the bridge girders expand with heat, the supports move on these rollers in order to compensate for the difference in length. Bearings of this kind can absorb the vertical forces affecting the bridge, but do not resist horizontal forces such as those caused by expansion movement as a result of temperature change, and they do not prevent the girders from deflecting either. For this reason they are called expansion bearings.

Fixed, articulated bearings

These supports are not on rollers and thus transfer horizontal as well as vertical forces. They are known as fixed bearings or simply articulations.

Restraint

What happens to the above-mentioned flagpole fixed into the ground? Its anchorage can transfer vertical and horizontal forces from the mast into the ground, and thus also prevent the mast from tipping over – a turning movement around the support. A support of this kind is called a restraint. › Fig. 8

We distinguish between three forms of support:

_ Simple supports can dissipate forces from one direction only. They slide and are articulated.

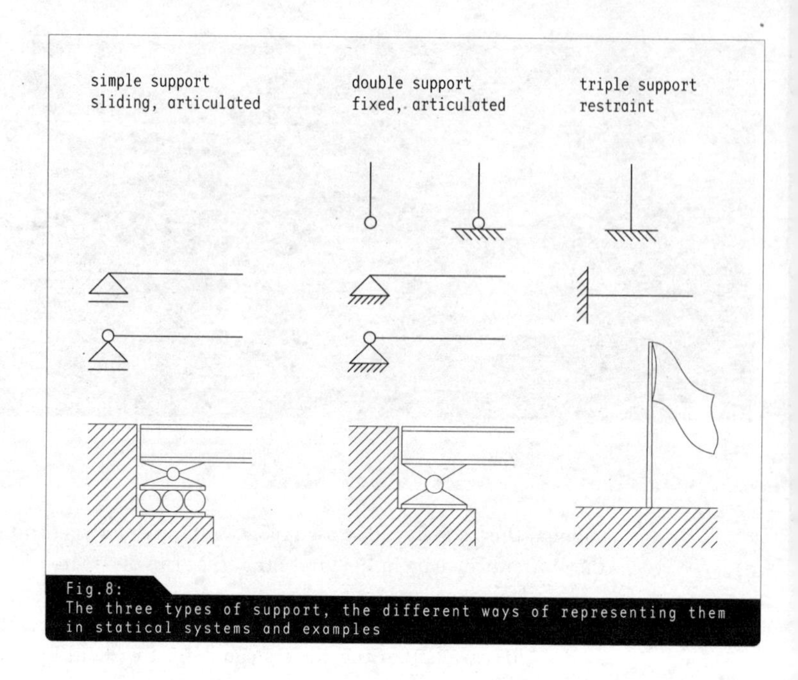

Fig.8:
The three types of support, the different ways of representing them in statical systems and examples

- Double supports can absorb forces from several directions. They are fixed and articulated.
- Restraints are triple supports and can absorb forces from different directions, as well as moments.

The correct choice of support is very important in construction, and must therefore be represented in statical systems. › Chapter Statical systems

Support forces

Let us assume that a beam is supported on a spiral spring rather than masonry. The spring is compressed by the load from the beam, thus creating a counter-force to the load that the beam exerts.

Support reaction

This force is called support reaction. › Fig. 9 If the beam does not move, the reaction force of the spring is exactly the same as the force exerted by the beam. Put simply: action equals reaction. › Fig. 10 It is not possible to see this in the masonry that usually provides support, but it is compressed just like the spring, so that it can generate the support reaction force.

When calculating a construction it is necessary to know the magnitude of the forces that the supports have to apply to support the structural element above them. The support forces are therefore always calculated

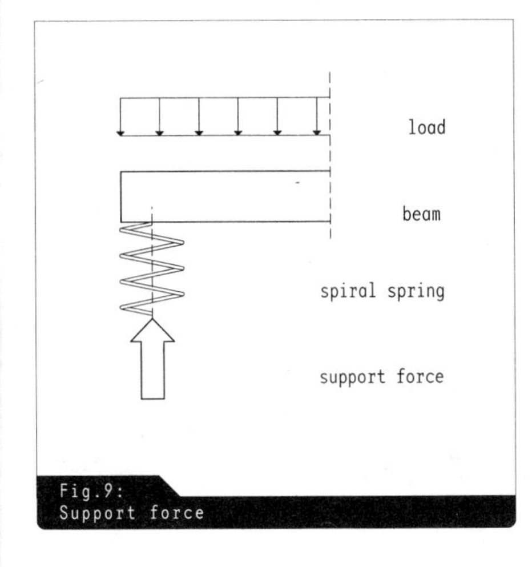

Fig.9:
Support force

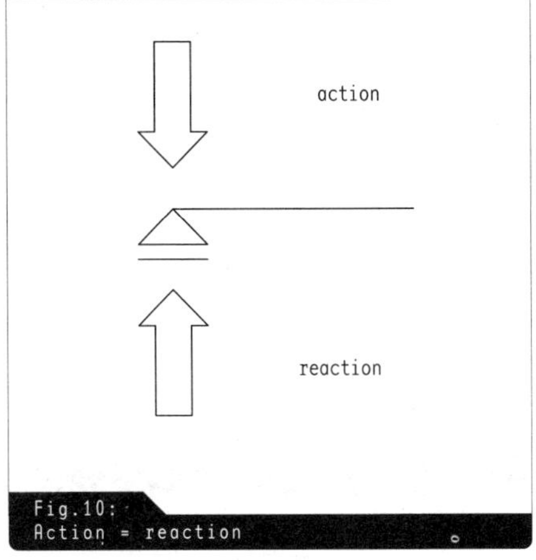

Fig.10:
Action = reaction

immediately after investigating the loads. Applying the above-mentioned law that action = reaction, it is possible to set up three theses for each structural element that make it possible to calculate the support forces. These three principles are the fundamental tools for statical calculations.

Conditions for equilibrium

They are also known as the three conditions for equilibrium: › Fig. 11

$$\sum V = 0$$

All vertical loads together are the same as all the vertical support reactions. This means: the sum of all vertical forces equals zero.

$$\sum H = 0$$

All horizontal loads together are the same as all the horizontal support reactions. This means: the sum of all horizontal forces equals zero.

$$\sum M_P = 0$$

If a support is considered at a support point P, all the forces turning clockwise around this point are the same as all the forces turning anticlockwise. This means: the sum of all moments around the given point equals zero. Here it should be noted that any force or load can be seen as a torsional force around a fixed point, so by definition the force times the lever arm length gives the size of the torsional force. › Chapter Forces

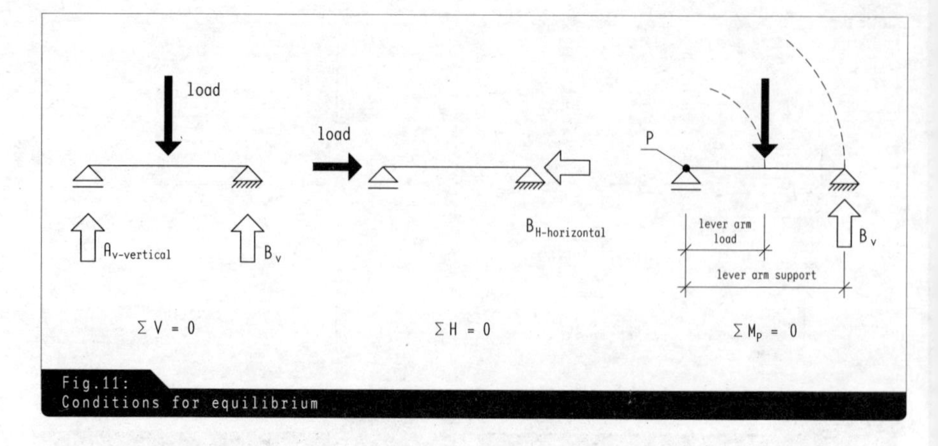

Fig.11:
Conditions for equilibrium

It is only by working out the sum of moments around a support that it becomes possible to calculate two different support forces. Because the torque centre is in a support, the support force does not have a lever there and equals zero in this equation. This means there is only one unknown in the calculation, and that is the other support force, which can be calculated easily.

So for the beams shown in Figure 11 with a single central load the sum of the torque around point P is as follows:

$$\sum \overset{\frown}{M}_A = 0 = A_v \cdot 0 + F \cdot l/2 - B_v \cdot l \rightarrow B_v = \frac{F \cdot l}{l \cdot 2} \rightarrow \quad B_v = F/2$$

Both supports dissipate half the central single load. This conclusion could have been reached without calculation in this case.

A rule of signs has to be decided for all calculations using the conditions for equilibrium. Rules of signs are not defined and so the statement must always be represented with an arrow. It shows the direction of the forces that are treated as positive. In this case turning to the right was treated as positive, so left-hand forces must be stated with a negative rule of signs.

INTERNAL FORCES

So far we have discussed only the forces impacting on a structural element and the forces the support generates as a reaction to them. These are called external forces, because the structural element itself has not yet

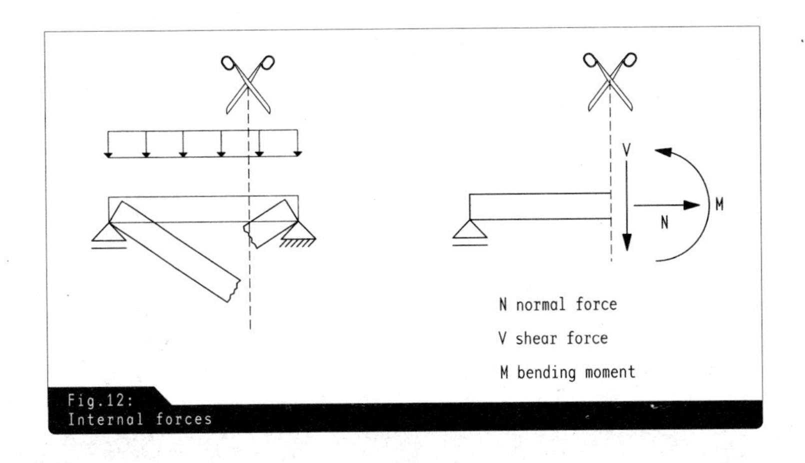

Fig.12:
Internal forces

been considered. But what is happening in the beam itself, or put another way, what forces are effective in the member?

To understand this, imagine a beam on two supports is cut through at a random point. What happens? It collapses and cannot support anything, not even itself. The crucial question now is what forces have to be effective at this cut face to prevent the beam from falling, or what forces are needed in order to achieve an internal equilibrium of forces.

Here the above-mentioned conditions for equilibrium are useful, as they apply equally to internal and external forces. It is assumed that the external forces acting from the plane of the cut to the end of the beam have to be as great as the internal forces that counteract the external ones at the plane of the cut. › Fig. 12

Internal focus

Just as the external forces are identified as vertical forces, horizontal forces and torque, internal forces are identified as normal forces, shear forces and bending moments, and their direction always relates to the structural element itself.

Normal force

A normal force is a force working longitudinally or in the direction of a structural element. As the first example illustrating a normal force we will take a rope hanging on a hook, with a weight attached to the rope. › Fig. 13 The weight is the load and the hook provides the support reaction. These would be the external forces.

Tensile force

Leaving out the weight of the rope itself, the same tensile force is effective at every point in the rope. Here it does not matter whether the rope is short or long. Therefore the same normal force is effective

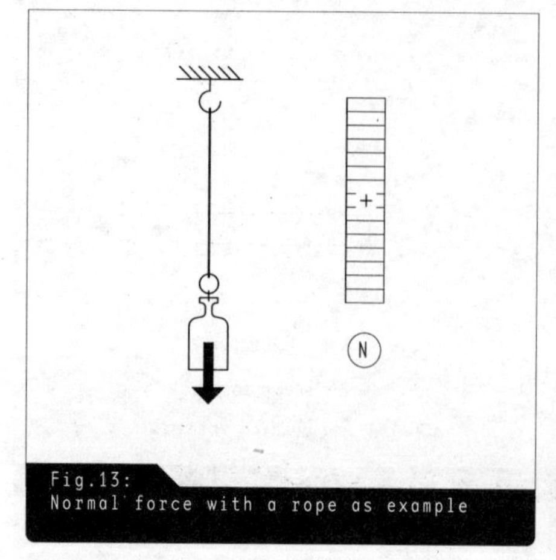

Fig.13:
Normal force with a rope as example

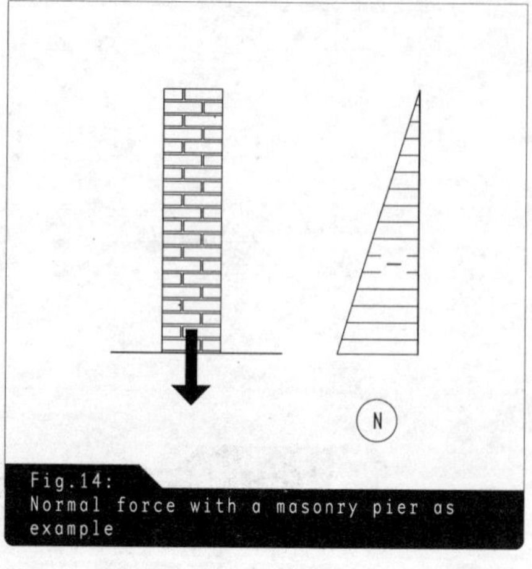

Fig.14:
Normal force with a masonry pier as example

at every point in the rope, and its magnitude is the weight of the weight attached.

Two longitudinal directions of forces were explained in the section Force action forms: pressure and tension. This is a tensile force.

We take a free-standing masonry pier for our next example. › Fig. 14 The pier's dead weight is the only load identified: masonry is a heavy material. It is easy to calculate the support reaction of the foundation at the base of the pier, as this must be the same as the weight of the pier. But what happens in the pier itself? The topmost stone takes no load from any other, so there is no normal force at this point. The second stone from the top takes the load of the one above it. So at the point of the second stone from the top there is a small normal force in the form of a compression force. This compression force becomes greater stone course by stone course to the bottom of the pier. That is to say, the normal force increases from the top to the bottom of the pier. › Chapter Forces

Compression force

The magnitude of the normal force can be demonstrated in diagrams, similarly to the representation of loads. The two examples show different flows of normal forces. In diagrams of this kind, tensile forces are noted with positive signs and compression forces with negative signs. › Figs 13 and 14

Sheer force

For external forces, a distinction is made between horizontal and vertical forces. Internal forces have the same relationship, but their direction relates to the system axis of the member in each case. Just as the longitudinally effective tensile and compression forces are defined as nor-

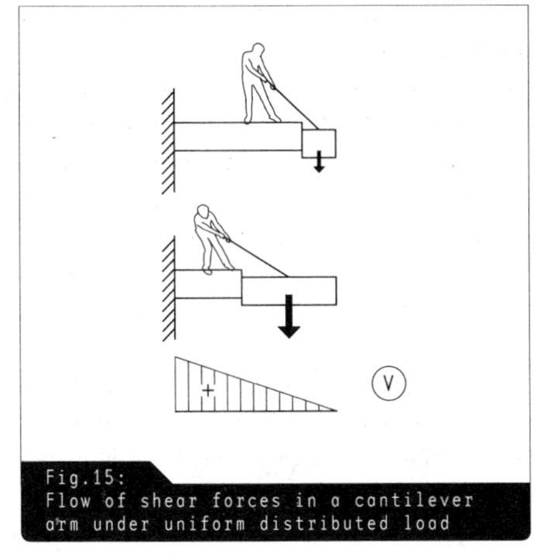

Fig.15:
Flow of shear forces in a cantilever arm under uniform distributed load

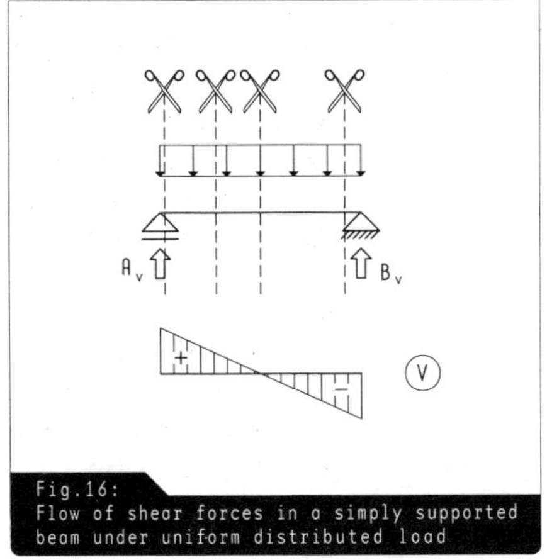

Fig.16:
Flow of shear forces in a simply supported beam under uniform distributed load

mal forces, all the forces working transversely to them are known as shear forces. They are not as easy to understand as normal forces, and must not be confused with bending, which is explained in the next chapter.

Cantilever arm

The effect of shear force will be explained taking a cantilever arm as example. Figure 15 shows a beam that is fixed into a wall at one end. This kind of beam is known as a cantilever arm. It could be part of a balcony, for example, and is loaded by its dead weight as a uniform distributed load. If this beam were cut through close to its end, the section cut off would fall because of this distributed load. The load works transversely to the axis of the member and thus produces the shear force. If a longer piece is cut off, more of the uniform distributed load has to be absorbed as a force transverse to the member axis at that point. Thus the shear force is greater at this point than at the previous one. The force would increase with every additional cut. The shear force thus increases from its free end towards the fixing point. So the support force at the point of fixing must be able to react to this shear force equally.

Simply supported beam

Figure 16 considers a beam on two supports, called a simply supported beam, with a uniform distributed load. The simplest thing to understand is the flow of shear forces when imagining one section after another cut off from left to right and considering what external forces are at work to the left of the cutting face.

The first interesting cutting face is just to the right of the support on the left. What happens in this section? The support force from the support is exerted upwards transversely to the member axis. Thus the shear force corresponds to the support force. But if a further cut is made to the right, part of the line force works in the other direction. This reduces the shear force in relation to the previous result.

Now a cut is made precisely in the middle. What forces are working transversely to the member from its left-hand end to the cutting plane? They are, first, the support force towards the top, and then the distributed load of the member section from the left-hand end to the centre. So half the distributed load of the entire member is effective. In a symmetrical system like this one it is easy to establish that each support dissipates half the distributed load. In this case the shear force in the middle of the beam equals zero.

If we now consider another cutting face to the right of this, an even greater proportion of the line force is effective. This means that the shear force becomes negative. At the cutting face just in front of the left-hand support almost the whole distributed load is working against the unchanged support force of the left-hand support. It is only the support force of the right-hand support that makes the result add up to zero again.

If the right-hand section of the beam had been considered instead of the left, the result would have been the same. So it does not matter which subsystem is considered, as the internal forces have to be in equilibrium at every point in the beam. This applies to all internal forces.

Bending moment

The effect of moments has already been discussed in the chapter External forces. Here all the effective forces were seen as turning around a fixed point. Their magnitude is defined as force times lever arm. › Chapters Forces and External forces, Support forces While the support forces were interesting for the external forces, the forces working in the run of the beams are important for determining the internal moments.

Bending

The internal moments cause the beam to bend. Bending is the key load for which many structural elements have to be dimensioned. When making statical calculations it is therefore necessary to know how great the bending moments need to be at any given point in the beam. This is shown in the moment gradient, which is thus an important aid for constructing members under bending load.

The direct link between internal moment and bending will again be explained below using a cantilever arm. How does the cantilever arm deform under a uniform distributed load? The load causes the beam to bend downwards. ›Fig. 17 Here, deformation through bending means that the beam has to become longer at the top and short at the bottom. This creates a tensile force at the stretched top side and a compression force at the squashed bottom side. These tensions counteract the load as an internal force.

Thus the bending itself creates the internal moments whose magnitude depends on the magnitude of the external forces and the length of

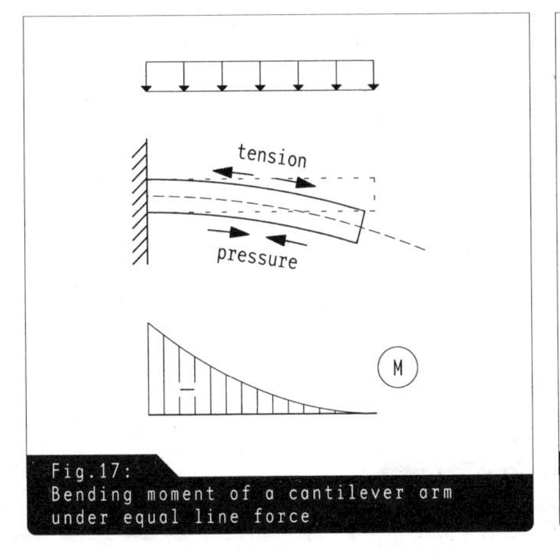

Fig. 17:
Bending moment of a cantilever arm under equal line force

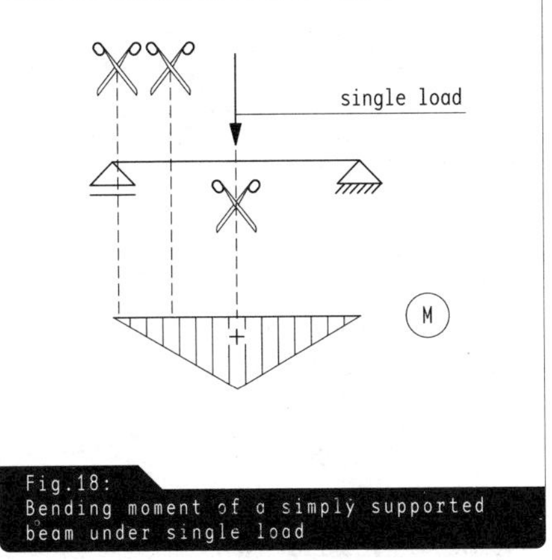

Fig. 18:
Bending moment of a simply supported beam under single load

their lever arm. For the cantilever beam there is low distributed load with a low lever arm force at the free end, and the moment is consequently small. But at the fixing point the full distributed load is effective, with a great lever arm force, so the moment is large. › Fig. 17

A simply supported beam under a single load will deflect downwards. Thus, in contrast to the cantilever arm described above, it will be stretched at the bottom and squashed at the top.

Here the bending is in a different direction from the previous example. How does this affect the flow of forces? Let us examine the beam from left to right.

The support force is effective at the right side of the left support, but it has no lever arm force, so the bending moment is zero. As the lever arm force increases with distance from the support, the moment increase is linear. This is the case up to the point where the single load is exerted. To the right of this, the single load works against the support force with increasing lever arm force, and the bending force diminishes until it becomes zero again at the second support. This test can be carried out as from the left-hand or the right-hand side as desired; the result is the same in each case. › Fig. 18

How does the flow of forces change if a uniform line force q is being exerted, rather than an individual load? A distributed load can be summed up as a resulting single load whose line of action lies at the centre of gravity of the distributed load. The magnitude of this resulting individual load is force per unit of length times its effective length.

$R = \frac{q \cdot l\ [kN \cdot m]}{m}$

To calculate the bending moments, these resultant individual loads and their lever arm lengths have to be established at the various cutting

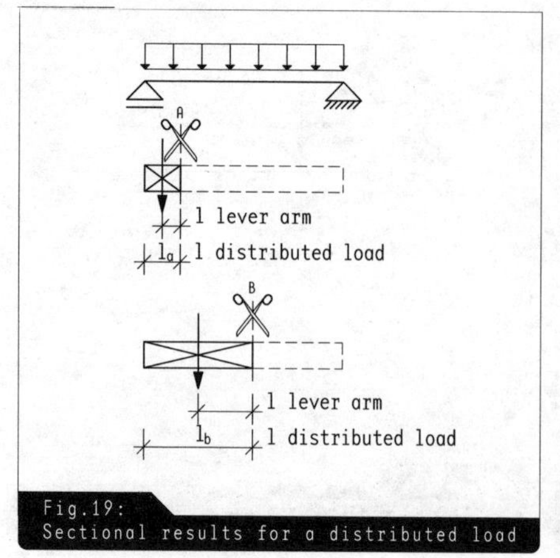

Fig.19: Sectional results for a distributed load

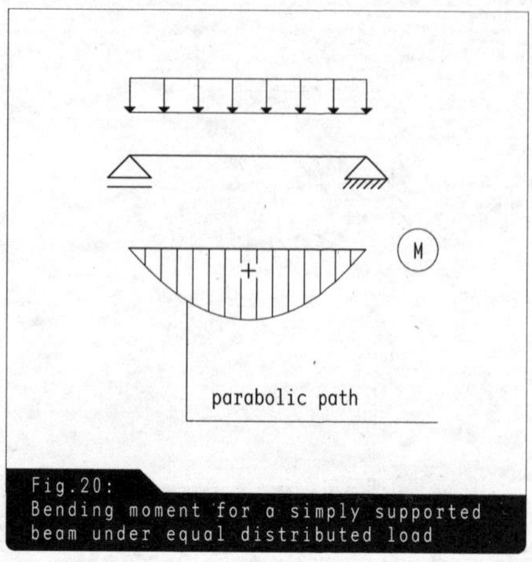

Fig.20: Bending moment for a simply supported beam under equal distributed load

planes. › Fig. 19 These act against the support force with increasing magnitude. The resultant moment gradient is a parabola, because the length as that of the distributed load and that of the lever arm goes into the calculation twice.

$$\text{Moment of a distributed load: } M_A = q \cdot l \cdot l/2 \rightarrow M_A = \frac{q \cdot l^2}{2}$$

The supports are important points for the moment gradient, and the bending force is again zero at both. How can this be explained? If we make a cut at the support and look in the direction of the support › Fig. 19 no force has a lever arm, because the member has no measurable length here; we are actually looking at a point. In general it can be said that bending requires a fixed beam cross section that can resist moments. But this does not apply if a cut is made in an articulated joint, which a hinged support is. A chain, for example, is an accumulation of articulated joints, and therefore cannot absorb bending. Hence we have an important principle: the bending moment in an articulated joint is zero. › Fig. 20

Maximum moment

In this example, the greatest moment is in the centre of the span. To bear loads, the beam must be able to resist this greatest moment. It is true in general that when dimensioning a structural element, the bending, the location and value of the maximum moment must be established.

When planning elaborate beams over large spans, the maximum moment is not the only important factor. It can be economical in terms of material to adapt the cross section of the beam to the moment gradient, in other words to shape the beam so that it is dimensioned precisely for the bending moment effective at every point. For this reason an architect should be able to determine the moment gradient in a beam qualitatively appropriately to the load.

Correspondence between the internal forces

The three different internal forces are introduced in the above paragraphs. When calculating loadbearing structures it is generally necessary to establish all three internal forces, so that a structural element can be dimensioned to deal with all three working together.

Shear force and moment correspond closely above and beyond this. The two forces, which result from the same load, permit inferences to be drawn from each other. For example, if no force is acting around a rod, the value of the shear force cannot change either, i.e. it is constant. But because moments are defined as force times lever arm, their changes of magnitude in an unloaded area are linear. If a force is acting at the place in question, the values of the resultant moment change in proportion to its distance from that place. Such interrelationships between shear force path and moment gradient are inevitable. › **Fig. 21**

The following correspondence in particular is important for statical calculations: if we compare the force flows, it becomes clear that the shear force has a zero crossover at the locations of the maximum moments. This turns out to be useful, because the location of the maximum moment can be ascertained from the flow of shear forces, and then has to be calculated at this point only. › **Figs 21 and 22**

And with a little experience, it is also possible to determine the anticipated shear force flow and moment gradient qualitatively. Figure 23 shows the appropriate force flows for some common load types.

1. If no force is effective in the member path, the flow of shear forces is constant and the moment gradient linear.
2. A single line creates a break in the flow of shear forces and a kink in the moment gradient.
3. For uniform distributed loads the flow of shear forces is a sloping straight line and the moment gradient a parabola.

\\Hint:
Signs for representing internal forces are fixed as follows:
Normal force: Pressure (–) is represented upwards, pressure (+) downwards.
Shear force: The positive shear force is drawn above, the negative below the system line.
Bending moment: The moments are drawn in the direction of their deflection, the positive moments downwards and the negative ones upwards. But these conventions should not be seen as definitive. For example, some countries represent the bending moments the other way round.

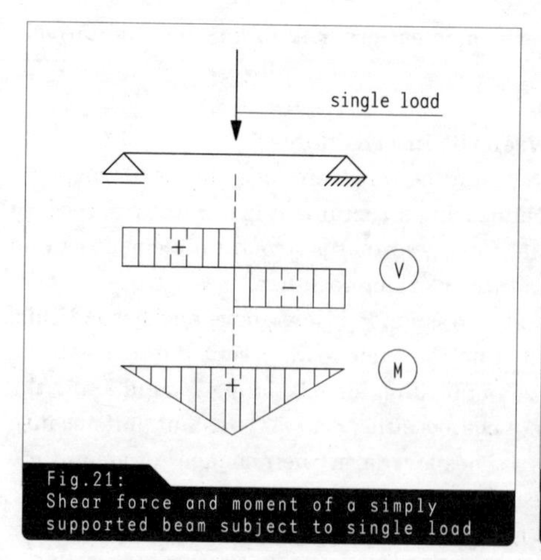

Fig.21:
Shear force and moment of a simply supported beam subject to single load

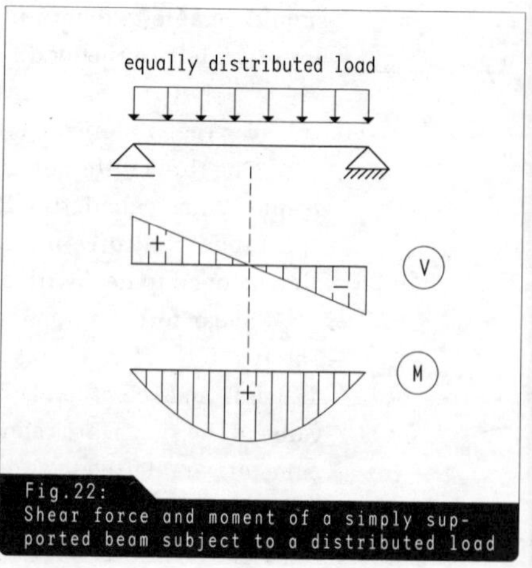

Fig.22:
Shear force and moment of a simply supported beam subject to a distributed load

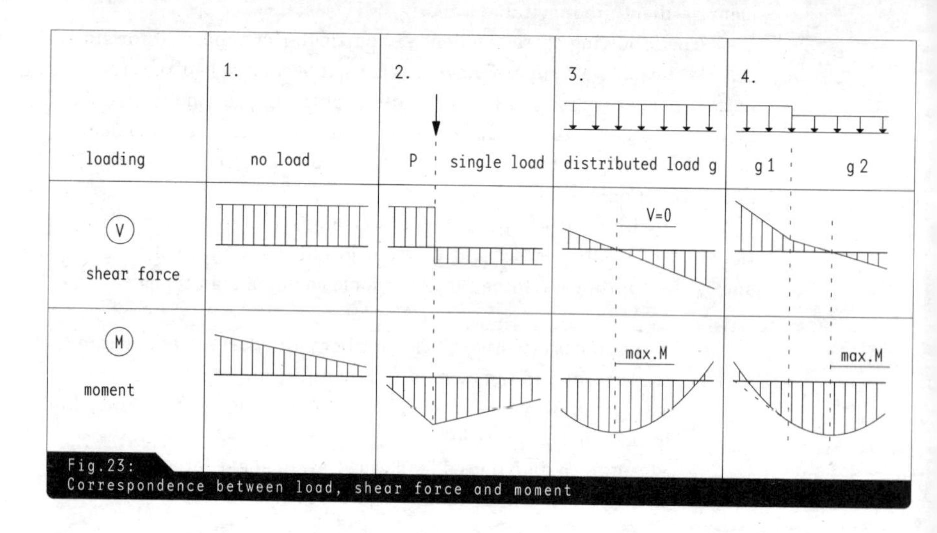

Fig.23:
Correspondence between load, shear force and moment

4. A break in the uniform distributed loads creates a kink in the flow of horizontal forces; in the moment gradient two parabolas with different inclines fit together with the result that they have the same tangents. Columns 1 and 2 in the table could be details from a system as in Figure 21, while Figure 22 can serve as an example for column 3.

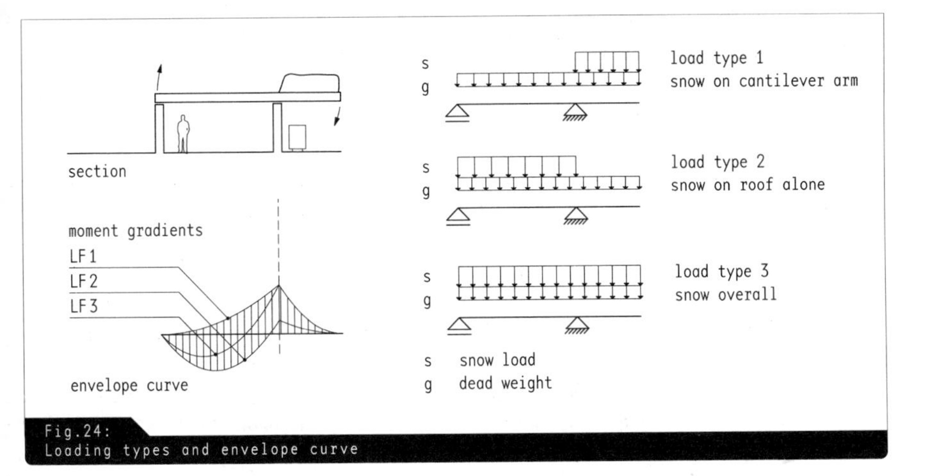

Fig.24:
Loading types and envelope curve

Loading types

In practice, many different loads overlap. They have to be added together in the calculation in order to dimension the structural elements for the maximum loading. But there are also cases where danger does not derive from the maximum load. Maximum values for internal forces, which are crucial to the dimensioning, are also possible for other load combinations. These different combinations of loads are called loading types.

Take the following example: a small workshop has a flat timber beam roof with a canopy protruding extensively on one side for the storage of material. A great deal of snow then falls in the winter. As the workshop is well heated and the roof poorly insulated, the snow on the roof melts, but not on the canopy, which has unheated space beneath it. The snow remains on the canopy alone. This load presents the risk of the workshop roof rising, and the roof beams above the wall next to the protruding section breaking. › Fig. 24

To avert these dangers, the structural engineer must calculate not only the snow load for the building as a whole, but the snow load on the canopy as well, as they each introduce different risks. The first step must be to establish what combinations or types of load are possible, and put them together. If the moment gradients of the different load types are then mapped onto each other, the possible extreme value for each point can then be read off from this diagram.

Envelope curve

This figure is also called an envelope curve. Its extremity shows the load type that is crucial for each point. Figure 24 shows that the greatest positive moment in the span occurs in load case 2, but the greatest negative moment in load types 1 and 3.

DIMENSIONING

A statical calculation runs similarly to the sequence in the above chapters. After the statical system has been established, the assumed loads are calculated and then the external forces, and after that the internal forces determined for the component parts of the building.

It would be wonderful if we could now simply work out the required cross section for the element. Unfortunately this is not as simple as it sounds, as all the parts of the building become part of the calculation themselves, as loads; in other words, all the structural elements have to be known in order to work out the assumed loads, so that their weight can be included. If the calculation then reveals that one of the structural elements assessed will not bear sufficient load, we have to start again from the beginning.

Even if this does not mean that all the work is invalidated, careful planning is clearly advantageous at this point: dimensions should be estimated in advance. This can be done with the aid of rough formulae. › **Appendix, Pre-dimensioning formulae**

Strength

After the forces coming into play have been determined, the load-bearing capacity of the structural elements is now of interest. This depends mainly on two aspects, the material and the cross section.

One of the first steps in designing a construction is to decide on the materials. Every building material has its advantages and disadvantages. The strength or resistance offered by different materials is particularly important for construction. Cable constructions, for example, resist tension but not compression, while masonry resists compression but not tension. Timber, steel and reinforced concrete constructions are compression- and tension-resistant, and also resist internal forces. › **Chapter External Forces, Force action forms**

It has already been established that compression and tension are both produced by bending. Consequently, only materials that resist both loads can be used when bending loads come into play (e.g. timber and steel members).

Tension

$$\sigma = \frac{F\,[kN]}{A\,[m^2]}$$

Materials also differ in their capacity to absorb forces. This capacity is expressed as the amount of force a material can absorb over a given area. The strength per area is expressed as tension σ.

Robert Hooke, 1635–1703

To understand the concept of resistance we must cite Hooke's Law, which states that tensions and extensions are proportional in the elastic field. What does this mean for building materials? Every material, whether it is wood, steel, reinforced concrete or masonry, is essentially elastic. If a structural element is loaded, tensions are created, causing the material to extend proportionately. So if a beam is loaded, it deflects, or sags. If the load is doubled, it deflects to twice the previous extent. If the load is reduced again the deflection also decreases.

This simple pattern applies only up to a certain point. If the tension becomes too great, the material no longer responds elastically, but plastically, i.e. permanent deformations occur. This is the point at which the structural element starts to be damaged. If it has to take a further load, it fails completely, although the failure differs in kind from material to material. The value indicating how much tension a material can absorb before it deforms plastically and fails is a purely material characteristic and has nothing to do with the geometry of the structural element. It is important for construction that the maximum admissible value is not reached at the highest possible load. The tensions a particular material can absorb are established under laboratory conditions, taking variations in material quality into consideration.

Admissible tension

The value established in this way is known as admissible tension and can be ascertained from tables. › Appendix, Literature

Strength classes

In addition, every material is available in different qualities with a variety of admissible tensions, and is assigned to a "strength class". For example, normal and high-strength concretes are distinguished by their strength class. The actual verification of a structural element's loadbearing capacity always works on the principle that the actual tensions must be less than the admissible tensions. The admissible tensions can be established from tables, but the main part of the work is in working out the actual tensions. If structural elements are loaded normally, working these tensions out is simple. The existing tension corresponds to the normal force per sectional area of the structural element. If the result shows that the existing tension is lower than the admissible tension, the structural element is correctly dimensioned. Unfortunately this simple verification is only rarely the deciding factor for dimensioning. Cables, which in fact can absorb only tensile forces, are dimensioned in this way, but in most cases the bending load is the key factor for dimensioning.

Moment resistance

Every proof of suitability for a structural element is based on the actual tensions being lower than the admissible tensions. This also applies to members subject to bending loads.

Stress distribution

When explaining the bending moment we stated that a bending member is subject to tensile stresses on one side and compression stresses on the other, but how great are these stresses, and how exactly are they distributed?

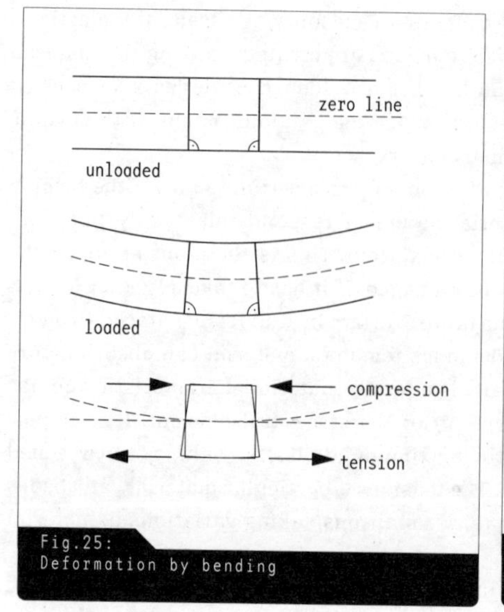

Fig.25:
Deformation by bending

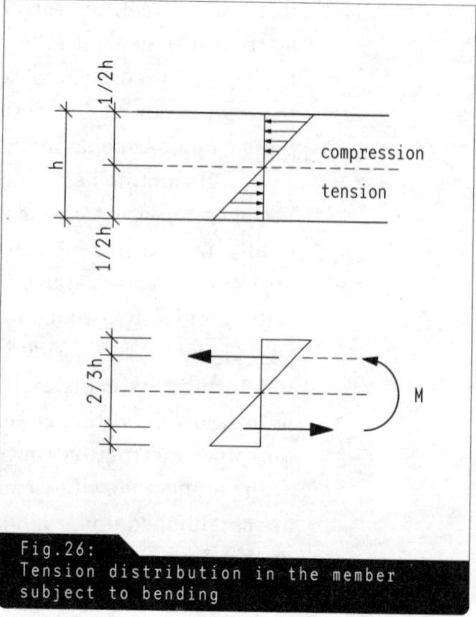

Fig.26:
Tension distribution in the member subject to bending

To find this out, consider an unloaded member marked transversely with straight lines. If it deflects when a load is applied, the marks incline towards each other in a trapezoid shape, but the lines remain straight. › Fig. 25 If the extensions and tensions are proportional, this results in a stress distribution that is also in straight lines from the tensile stress at the lower edge via the middle level, which is tension-free, to the compression stress at the upper edge.

Neutral tension plane

As can be seen in Figure 26, the compression and tension distributions each form a triangle. These triangular tensions can each be summed up in a resulting tension at the centre of gravity of the triangle, with a distance from each other of 2/3 of the height of the cross section. This length represents the lever arm of the internal moments that counteract the loads and are thus responsible for loadbearing capacity. So the greater the height of the member, the longer the lever arm of the internal tensions, and the greater the stability of the member.

Thus, the length of this lever arm is the key to resistance to bending, but the section width is also important. This section resistance is expressed as moment resistance. Moment resistance is a value relating to the geometry of a member and not to its material.

For example, the rectangular cross section customary in timber construction produces the moment resistance W with a magnitude of $W = w \cdot h^2/6$.

It is worth looking more closely at this formula: the height h is squared, while the width w is simply entered as a factor. An upright rectangular section has a higher loadbearing capacity than a square one, or a horizontal rectangle. Expressed more precisely, doubling the width of a profile doubles the loadbearing capacity, but doubling its height multiplies it by four.

The moment resistance for dimensioning a simply rectangular cross section can be established by using the formula above. For other cross sections, for all steel profiles, for example, the dimensioning is more complicated. For this reason, moment resistance values are always given in sets of tables. › Appendix, Literature

The term moment resistance contains the word moment. In contrast with the concept of moment or torque explained in the chapter Forces, moment resistance refers not to an individual force with a particular lever arm, but to area elements and their lever arm around the tension zero line, as we are considering a cross section area. › Fig. 26 Like moment of inertia, explained in the following chapter, moment resistance is also defined as an area moment.

Moment of inertia

Moment of inertia is best explained by its effect. Moment resistance expresses a member profile's resistance to bending moments, while the moment of inertia relates to its deflection. It describes the rigidity of a cross section.

Like moment resistance, moment of inertia is based on the distribution of tension in a cross section subject to bending. Here the severely compressed and extended areas at the edge are more effective than those in the zero tension line area. But the distance of the area elements from the zero line is more important for moment of inertia than moment resistance.

The moment of inertia is the sum of all area elements in the cross section multiplied by the square of their distance from the zero line.

The moment of inertia for rectangular cross sections is calculated using the formula $I = w \cdot h^3 / 12$. So here the cross section height is raised to the power of three, which means that deflection is reduced by one eighth if the member height is doubled and the width remains the same.

Deflection

Moment of inertia can be used to calculate the expected deflection of a member. Even though the priority is to dimension structural elements for loadbearing capacity through moment resistance, evidence is additionally needed that an admissible maximum deflection is not being exceeded.

Shear stress

Let us take the following example to explain shear stress: two planks are laid one on top of the other as a simply supported beam, and a load

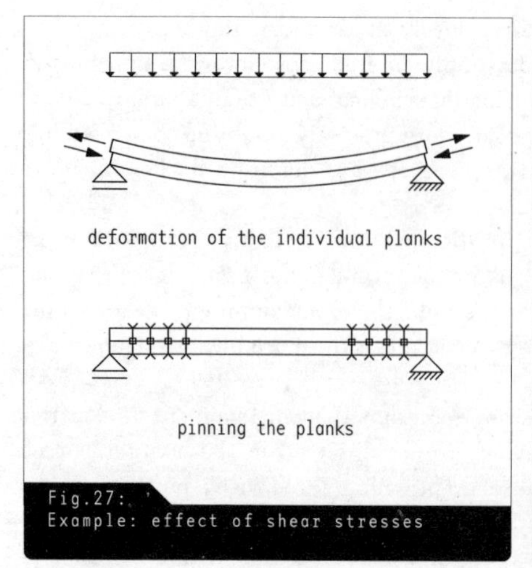

Fig.27:
Example: effect of shear stresses

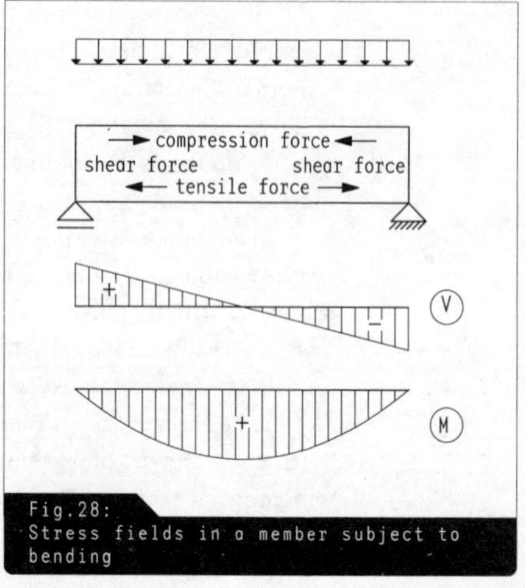

Fig.28:
Stress fields in a member subject to bending

is then exerted on them. Both planks will deflect under the load and shift in relation to each other. › Figs 27 and 28 They should be fastened together to increase their loadbearing capacity, as a tall cross section has a higher loadbearing capacity than two planks on top of one another with the same height. › Chapter Dimensioning, Moment resistance and Moment of inertia What is the best thing to do?

One possibility could be to drill through the unloaded planks and fasten them together with bolts and dowel pins.

But now we have to ask what stresses these dowel pins are actually expected to absorb and how they come into play. The answer to the first part of the question is simple. Shear stresses are responsible for the planks' shift in position. The simplest way to explain where these stresses come from is by means of the members shown in Figure 28.

In this member subject to a uniformly distributed load the greatest tensile stress is in the middle of the span on its lower side and the greatest compression stress on its upper side. These stresses reduce as the bending moment decreases towards the supports, but what happens to the stresses that cannot just simply disappear? The compression and tensile stresses cancel each other out towards the supports and this happens as a result of the shear stresses that increase as the bending stresses decrease.

The bending stresses can be discerned in the moment gradient, while the shear stresses are proportional to the shear forces and in the case of a member subject to a simple and uniform load the shear forces increase towards the supports. › Chapter Internal forces, Shear force

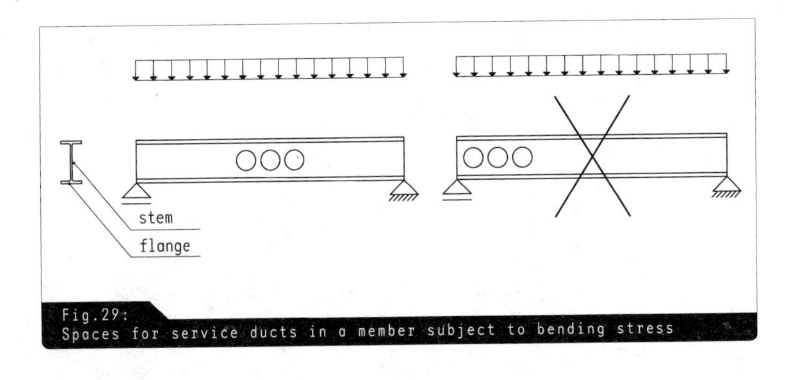

Fig.29:
Spaces for service ducts in a member subject to bending stress

It is essentially true of members of this kind that the tensions resulting from bending in the centre of the span are greatest at the top and bottom of the section › Chapter Internal forces, Bending moment and the maximum shear stresses are at the supports. Timber, for example, is a material that is sensitive to shear stresses. In timber structures it is often necessary to reinforce members to absorb shear stresses at the supports.

I-beam section

Another example is offered here to clarify shear stresses further. The usual steel sections, such as I-beam sections, are designed so that the flanges can absorb the bending compression and bending tension, while the stem absorbs the shear stresses.

For example, if an architect wants holes for wiring or pipework to be cut in a simply supported beam, this does not present a problem in the middle of the beam, as the forces in play tend to be small there, while the two flanges are fully loaded with bending compression and bending tension. But holes should not be drilled near the supports, because here the stem is heavily loaded with shear stresses. › Fig. 29

›✎

✎

\\Tip:
The cabling and ducts for power, water, sewage disposal and, above all, ventilation can have a crucial effect on the design of a loadbearing structure. They should be fixed at an early stage and agreed with the structural engineer. Essentially, loadbearing structure and service pipes and ducts should be planned so that as few crossing points are created as possible.

CANTILEVER ARM, SIMPLY SUPPORTED BEAM, SIMPLY SUPPORTED BEAM WITH CANTILEVER ARM

Loads and forces were explained in the first chapter using the cantilever arm and the simply supported beam as examples. These two load-bearing systems form the basis for most of the more highly developed and more complex systems. It is worth recapitulating their advantages and disadvantages.

Cantilever arm

A cantilever arm can be compared with a long lever used to lift heavy loads. Consequently the leverage acting at the anchor point is the biggest problem. As can be seen in Figure 30, this is the point of maximum torque and maximum shear force, and the anchor point has to absorb both. This is scarcely feasible in timber construction, as no nailed or screwed joint could do the job unless the anchor point were long enough. But an anchor point in masonry is easily possible, although the danger remains that the long lever could lift masonry even if there is not enough of it above the anchor point. If we look at the moment and shear force gradients, it is clear that a cantilever arm subject to a uniformly distributed load has to be dimensioned for the area around its anchor point, but is thus overdimensioned for the rest of its length. It therefore makes sense, saves material and is customary for a cantilever member to have its height reduced from the anchor point to the free end to correspond with the internal force gradients.

Simply supported beam

The simply supported beam is probably the most common loadbearing system, and it is worth looking at carefully again here. A simple timber, steel or even concrete section with a consistent cross section is normally

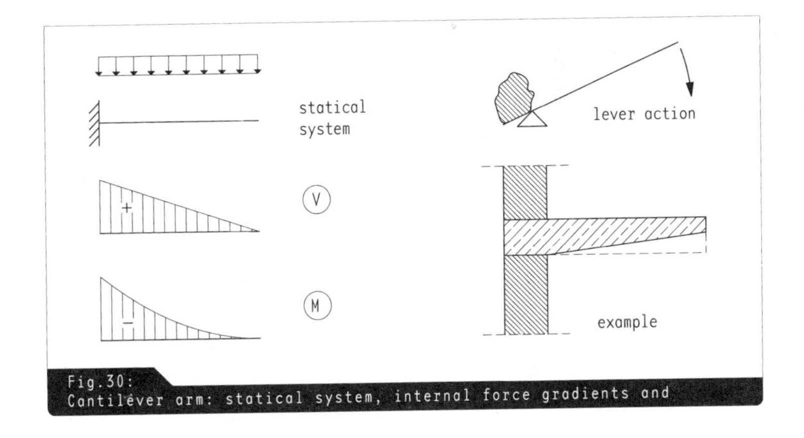

Fig.30:
Cantilever arm: statical system, internal force gradients and

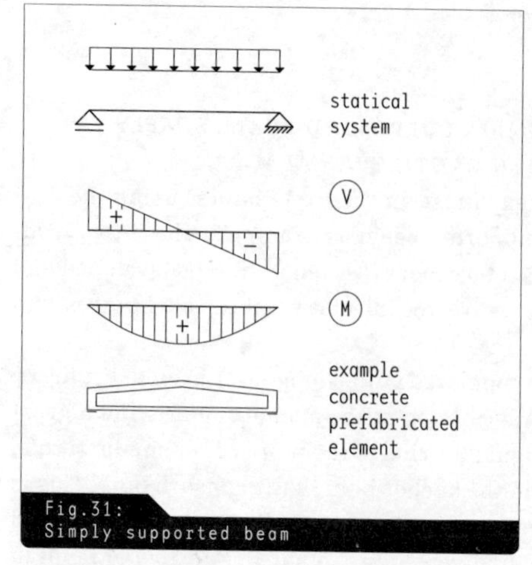

Fig.31:
Simply supported beam

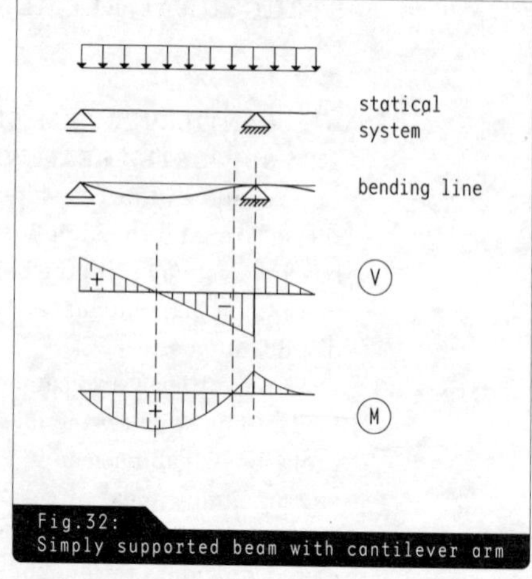

Fig.32:
Simply supported beam with cantilever arm

used, as these are easy to produce and cheap, and also have the advantage of offering a flat surface at the top and the bottom. But actually most simply supported beams are only fully exploited at one point, in the middle, the point of maximum moment. So it can make sense to adapt the beam to the moment gradient and make it higher in the middle than at the supports.
› **Fig. 31 and Chapter Internal forces, Bending moment**

Wood is a natural product that can absorb considerably fewer forces transversely than it can longitudinally to its grain, and it is thus sensitive to shear forces. In the best cases a timber beam is thus fully exploited at three points, in the middle because of the bending moment and at both ends because of the shear forces. › **Chapter Dimensioning, Shear stress**

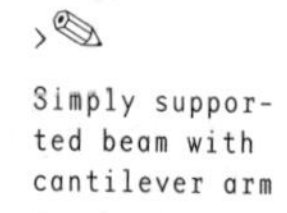

Simply supported beam with cantilever arm

The simply supported beam with cantilever arm is a very useful system from the point of view of loadbearing theory. It could be said that as a combination of the two previous ones it compensates for the disadvantages in each case. The problem with the cantilever arm is its anchor point. But in this system the length of the anchor point, i.e. the span length, is usually greater than the protruding section itself, and thus unproblematic. The key factor for this beam is what happens above the support in the cantilever arm. Here, the cantilever section has its maximum moment, with a negative value. › **Fig. 32**

A simply supported beam has its maximum positive moment in the middle of the span while at the support it is approaching zero. How do these two lines fit together? This becomes clear if we imagine a member of this kind deflecting when loaded. The cantilever arm would hang down and the member would also sag in the span. It hangs in at the articulated end support, but curves over the other one and lies horizontally over the support.

Bending line

This means that the inflection point of the "bending line" shifts from the support into the span. › Fig. 32

This is also reflected by the moment gradient. Corresponding with the protruding section, the negative maximum lies above the support.

Moment at support

Midspan moment

A negative above a support is called a moment at support. It is not relieved in the span before a "midspan moment" comes into being. This is the term for the positive moment in the area between the supports. Because of the moment at support it is smaller than in the case of a pure simply supported member. The member in the span is thus relieved by the cantilever arm. This means that a member of this kind can be smaller than a simply supported beam over the same span width.

CONTINUOUS BEAM

Continuous beams extend over several spans. They are defined precisely according to the number of these spans. A two-span member has three supports, a three-span member four, and so on. Such systems are a logical extension of the situation explained above. As in the simply supported beam with cantilever arm, a moment at support is created above a central support, and this reduces the bending moments in the spans. Here, the inflection points in the bending line correspond with the zero points in the moment line, although the bending line and the moment line do not have the same form. But the form of the bending line indicates the moment gradient. › Fig. 33 and Chapter Internal forces, Bending moment

So the advantage of continuous members is that they reduce the span moments through the moments at support over the supports. Lower span moments means that the members can be smaller.

\\Tip:
Squared timber sections used as beams should be neither too slender nor too wide. Sizes with side ratios between 2/3 and 1/3 make sense.
Building tables give timber sections. The section sizes quoted in these tables are usually available from stock in the timber trade, so they do not need to be specially cut to size, which would make additional work for the carpenter.

\\Hint:
A positive moment or a midspan moment signifies tensile stress on the bottom side and compression on the top side of the member section. A negative moment or a moment at support creates compression on the top side and pressure on the bottom side (see Chapter Internal forces, Bending moment).

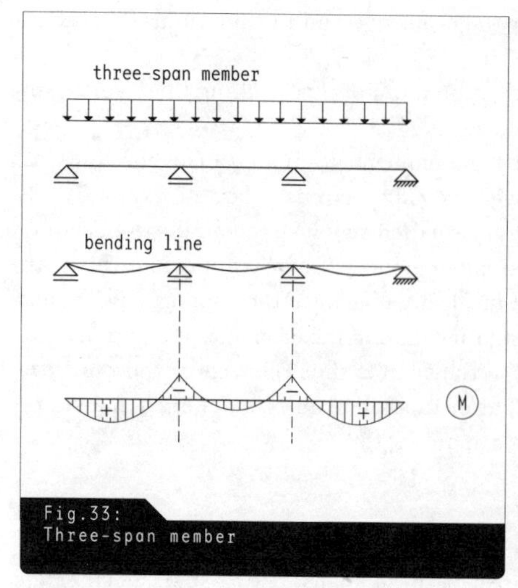

Fig.33:
Three-span member

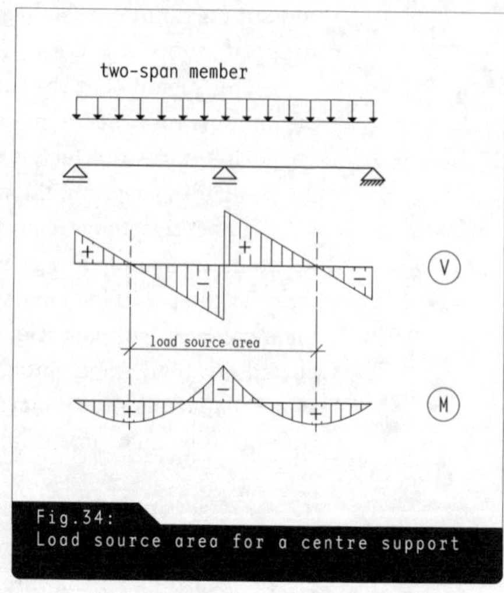

Fig.34:
Load source area for a centre support

Effect of continuity

The effect of continuity thus makes significant material savings possible.

Without closer consideration it might well seem that a central support has to bear exactly twice as much load as the peripheral supports. But this is not the case. The threshold of the load source area is not in the middle of the span, but at the point where the shear force is zero and the span moment at its maximum. Thus, a central support takes more than twice the load at a peripheral support. › Fig. 34

ARTICULATED BEAM

Another possibility emerges when looking at the moment gradient of the continuous member: the individual beams can be fitted together at the points of zero moment. This sustains the effect of continuity and, above all, a point of zero moment means that there is no bending at this point. So if a beam joint is planned for this point, the moment gradient does not change when compared to the continuous beam. In timber construction a beam joint is an articulation, as is the case with almost every nodal point, and the bending moment at an articulated point is inevitably zero. › Figs 35 and 36

Point of zero moment

Statical determination

An added articulated joint affects the system in a further way. Continuous beams and articulated beams differ in one essential quality. What happens to a continuous beam if one of the supports is lowered for some reason? The beam will have to bend in order to remain supported by all of them. This creates stresses in the structural member. If this were to happen

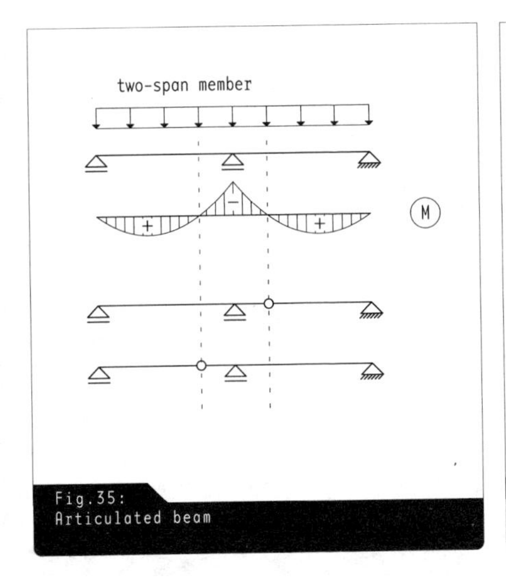

Fig.35:
Articulated beam

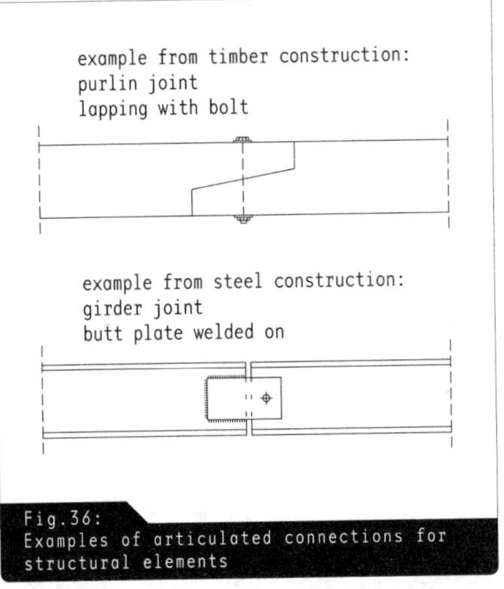

Fig.36:
Examples of articulated connections for structural elements

to an articulated beam, there would be no stresses in the section, because of the articulated, sliding nature of the support system.

Statically undetermined

Loadbearing structures in which stresses would be created if a support were to be moved are called statically undetermined, and if this does not happen, they are called statically determined. › Fig. 37

Statically determined

For example, cantilever arms and simply supported beams, to which this distinction applies as well, also prove to be statically determined systems. More loadbearing systems will be explained in the following chapters that can be statically determined or statically undetermined. Statical determination always depends on the number and nature of the supports, and the number of articulation points. Adding articulations can turn a statically undetermined system into a statically determined one. But care is needed here, as a superfluous addition would render the system unstable.

Three-span member

Closer consideration of a three-span member with a uniformly distributed load shows that it would need two articulated joints in order to raise or lower each joint without causing stresses. In other words, two articulated joints are needed to make it into a statically determined system. Four points of zero moment show in the moment gradient. There are thus several possibilities for arranging the joints within these points. › Fig. 38

What is the difference between statically determined and undetermined systems in practice? Statically undetermined systems offer a somewhat higher degree of safety based on the distinctions identified above. If, for example, one support for a continuous member should fail, there is

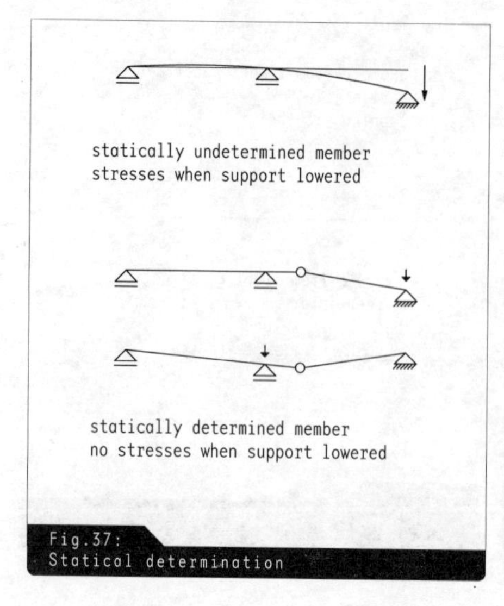

Fig.37:
Statical determination

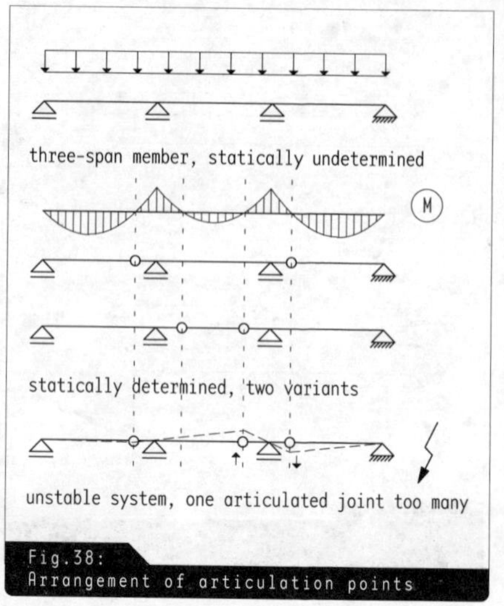

Fig.38:
Arrangement of articulation points

a chance that the structural element will not collapse because the member is still supported by the remaining ones. In a statically determined system, such as a simply supported beam, this would not be the case. In addition, statically undetermined systems cannot be calculated using the three conditions for equilibrium. More elaborate calculation methods are needed here.

TRUSSED BEAM

The span width is the most important criterion for choosing a particular loadbearing system. In any construction, it is possible to assign a sensible point for a width at which the span can still function, but will become inefficient if that point is exceeded. For example, this point is reached at approx. 5–6 m for the efficient use of single timber beams. Further measures are needed for wider spans: for example, if it is not possible to place a support underneath, a brace can be inserted instead, › Fig. 39 which will dissipate its load to the supports via a truss. The truss pushes the brace upwards, like a drawn bow, and thus works like a support, even though it does not touch the ground. This system is called a trussed member or beam.

It is also possible to truss a beam doubly or triply. › Fig. 40 This increases the span even further, though the forces in the structural members increase correspondingly. How are the individual parts of the trussed member loaded? The brace is compression loaded, as it is supporting the beam. The truss, which is usually made of steel rods, is subject to a tensile

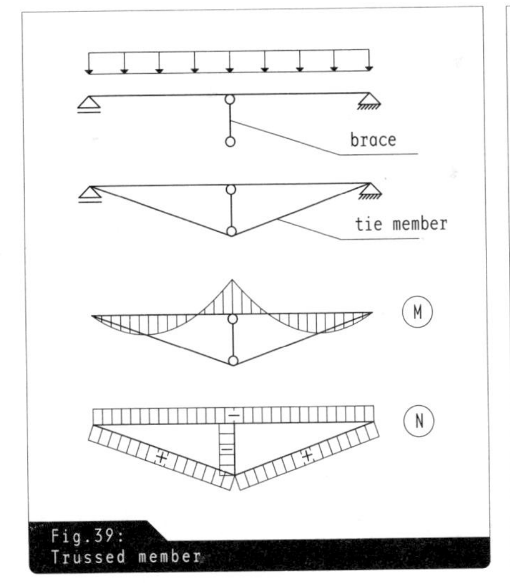

Fig.39: Trussed member

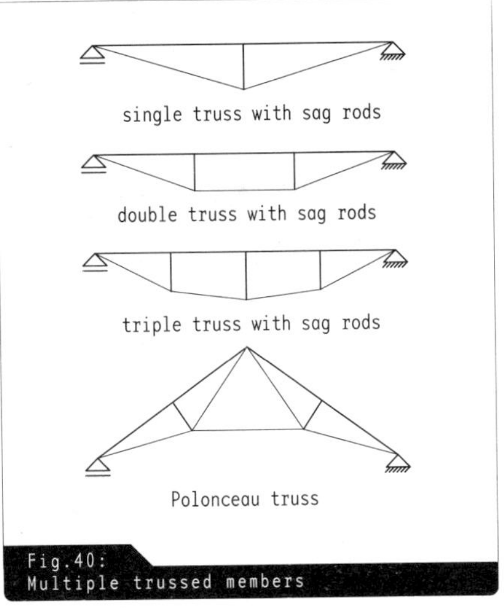

Fig.40: Multiple trussed members

force and the beam, which was originally subject only to a bending load additionally acquires compression as a counterforce to this tensile force.

Jean B.-C. Polonceau, 1813–1859

More complicated systems can be constructed with the aid of braces and trusses. One example is the Polonceau truss, named after its inventor. › Fig. 40

It is possible to sum up what was achieved by the trussed member as follows: the simple beam becomes a complex system that deals with the loads not just by absorbing bending moments, but by dissipating them as compression and tension in various structural elements set at a great distance from each other. The upper member is no longer subject mainly to bending forces, but also to compression, and the truss dissipates the tension forces. When explaining the bending moment, we talked about the lever arm of the compression-loaded cross section parts as opposed to the compression-loaded ones. This lever arm is clearly enlarged here. Hence, such systems are considerably more efficient. › Chapter Dimensioning, Moment resistance

LATTICE

A trussed girder with more than three struts makes less sense for a number of reasons. But if the struts are supported individually in every section, this produces a new system that can cope with considerably larger spans. It is known as a trussed, lattice or skeleton girder. In these lattice girders the tension-loaded sections are usually made up not of cables or

Fig.41:
Steel lattice constructions

bars, but of timber or steel sections. Lattices are efficient systems that are very common, and can be adapted to fit the requirements of a particular situation. They can be realized in almost any material, and the bars can be arranged in very varied ways. › Fig. 42

Tension diagonal

In the examples discussed so far, the diagonals have been realized as tension-loaded bars, corresponding to the truss below them. But it is equally possible to install the bars exactly the opposite way round.

Compression diagonal

They are then compression loaded. To identify the direction of forces in the diagonals, it helps to ask whether they are loaded with compression forces, like an arch, or with tension forces, like a sagging cable.

Alternate diagonals

It is also possible to construct lattice trusses with diagonal members alone, in alternate directions. The look of the truss does not change in the

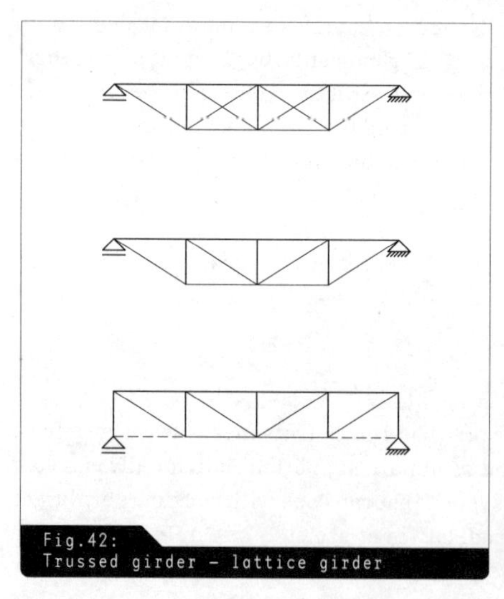

Fig.42:
Trussed girder – lattice girder

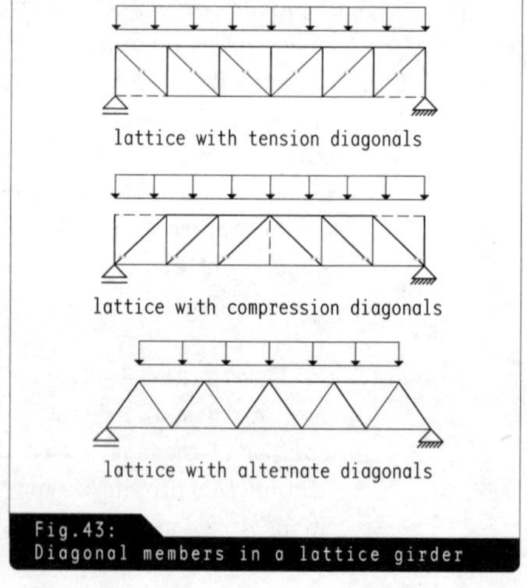

Fig.43:
Diagonal members in a lattice girder

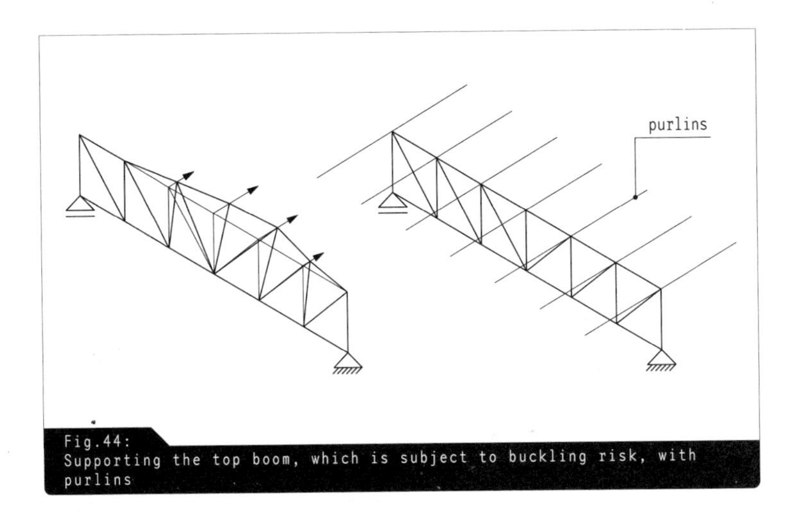

Fig. 44:
Supporting the top boom, which is subject to buckling risk, with purlins

middle. The bars in one direction are compression loaded and the others tension loaded, and in the middle of the lattice structure the sections change their load, while their arrangement remains the same.

Unstrained members

There are individual bars in lattice girders that on closer consideration are not directly involved in dissipating the load. They are neither tension nor compression diagonals, and are thus called unstrained members. Nevertheless, they cannot usually be omitted, because they are needed for structural reasons. This can mean that they complete the outlines of the system or hold it in position. In the figures, compression members are drawn as thick lines, tension members as thin lines and unstrained members as dashed lines. › Fig. 43

Top boom

The height and length of a lattice girder is calculated according to the span. But its width depends only on the girder section selected in each case, and is usually very narrow in comparison with the overall length. For this reason the compression loaded upper section, called the top boom, is at risk of buckling. › Fig. 44

This problem can be solved in a variety of ways.

Purlins

The top boom can be fixed to a ceiling above it or to longitudinal ceiling beams, and thus prevented from moving out of place; or it can be constructed as a buckleproof girder in its own right.

Three boom truss

If a second boom is added to the top boom, which is in danger of buckling, and a kind of lattice is constructed with diagonal braces between the two, this produces a rigid support element in both directions, called a three-boom truss. › Fig. 45

›✎

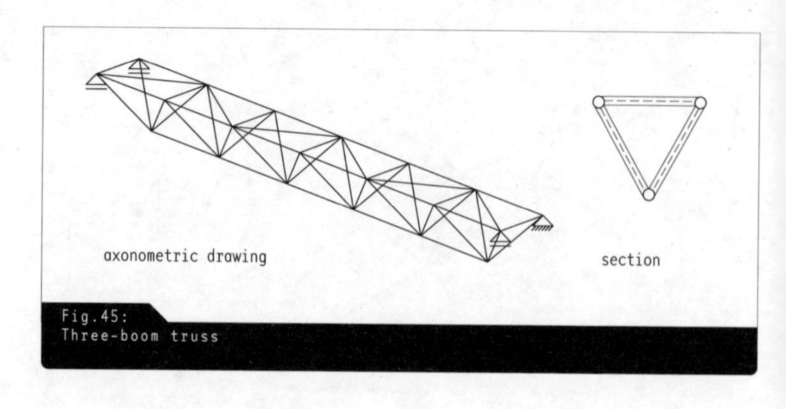

Fig.45:
Three-boom truss

SLAB

Timber or steel constructions are almost always directional systems, i.e. the bar-shaped sections means that loads are always dissipated in a particular direction. Concrete, however, makes it possible to created statically non-directional, flat structural components.

Reinforced concrete

Fundamentally, the following applies to the loadbearing properties of reinforced concrete: as artificial stone made of cement, water and aggregates such as gravel or chippings, concrete is very good at absorbing compression forces, but like masonry does not absorb tension forces well. It is therefore usually combined with steel.

Reinforcement

In this material structure, the concrete absorbs the compression forces and the steel the tension forces. The previous chapters on the various support types have already explained where tension forces occur in structural components. This is precisely where the reinforcing elements are placed in reinforced concrete, i.e. bar steel is cast in. For slabs, this is usually needed on the undersides and in peripheral areas. If a reinforced concrete slab is used like a continuous girder, steel reinforcement is also built in on the top side. When concreting a floor slab, steel bars are usually inserted in the form of mats welded crosswise, enclosed by concrete on all

\\Tip:
The nodes in lattice girders should be constructed so that the sections – or more precisely, their centre lines – meet precisely at one point. This avoids generating forces that twist the node and would subject it to additional loading.

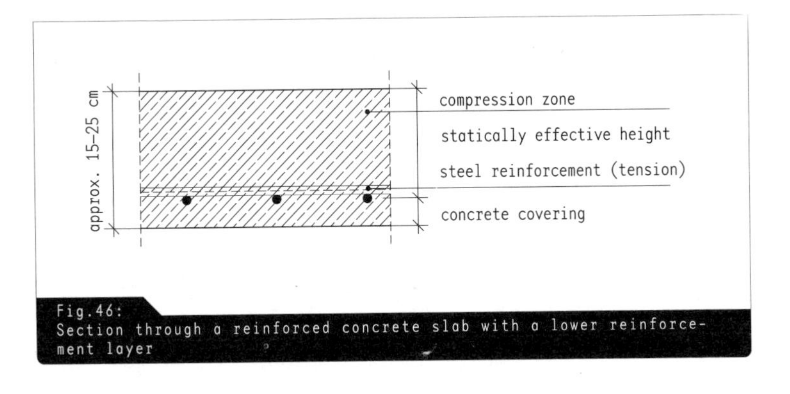

Fig. 46:
Section through a reinforced concrete slab with a lower reinforcement layer

sides to be able to support the load compositely. The thickness needed for a concrete floor depends on the span, and is usually 15–25 cm. › Fig. 46

Concrete slabs are almost the only structural components that can be non-directional. Over a square space, a concrete floor can dissipate its loads to all four walls at the same time. But if the floor is rectangular the loads will be dissipated via the short span in the first place, because if the deflection is even it will be more heavily loaded than the long span, thus producing greater tensions. For a concrete floor that is twice as long as it is broad, the proportion of loading absorbed by the long span is scarcely significant. But the reinforcement is not simply fitted to relate to the main direction in which the load is borne. A slab is always reinforced transversely as well, as the flat effect brings its own advantages. This means, for example, that point loads are better distributed, and the forces in the floor remain lower.

The longer the span, the thicker the floor will need to be. But if a floor is thicker than 25 cm, its dead weight becomes so large that it is scarcely viable any longer as a solid flat floor. Strictly speaking, only the upper edge of the floor is effectively involved in dissipating the compression load, and the steel bars dissipate the tension stresses. The rest of the structure is actually just a link, or a filler.

Ribbed floor

If floors are very thick, it makes sense to reduce their dead weight by omitting areas from the lower edge to the effective upper zone. The reinforcement is then placed mainly in ribs, which lie very close to each other. A ribbed floor can thus accommodate much wider spans than a flat one.

Binding joists

Another approach to bridging large spans involves using binding joists. Unlike ribs, binding joists are not seen as part of the floor area, but as beams on which the flat floors will be laid. › Fig. 47 and Fig. 69, p. 63

Slab beams

For concrete structures cast on the building site (in-situ concrete structures), binding joists best use the advantages of the monolithic

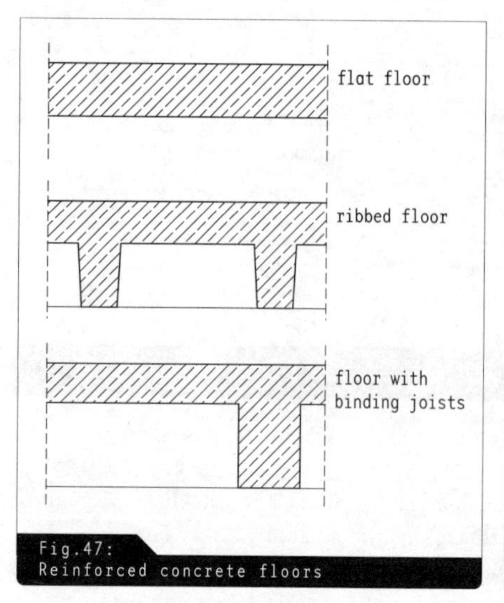

Fig.47:
Reinforced concrete floors

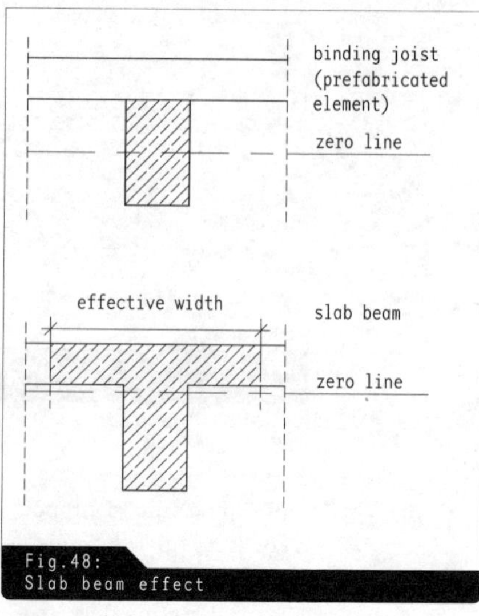

Fig.48:
Slab beam effect

construction method. Here, "monolithic" means that all the in-situ concrete elements, even if they have been concreted at different stages, work as a continuous structure. So binding joists exploit not only the static height to the bottom edge of the slab, but also its thickness. Furthermore, the part of the slab by the beam on both sides enlarges the compression zone. In such cases, the term slab beam is used. › Fig. 48

COLUMN

Unlike horizontal loadbearing elements, columns are subject to hardly any bending load, but primarily to normal forces. A very narrow column cross section would suffice for dissipating normal forces if it were not for the danger that the column might sag sideways and fail.

Buckling

Slender columns run the risk of buckling, but the magnitude of this risk depends on various factors. The important features for a column are the loads, the material, and how slender it is. The Swiss mathematician Leonhard Euler (1707–1783) established how the way columns are fixed at the top and bottom affects their buckling properties, and identified four different cases, which are named after him.

Euler cases

The Euler cases set out four ways in which columns can be braced or provided with articulated joints. When buckling, columns adopt the form of a sinus curve. The way the columns are attached affects the length of this sinus curve, or the distance between their inflection points, which is important in its turn for the stability of the column. The length of the

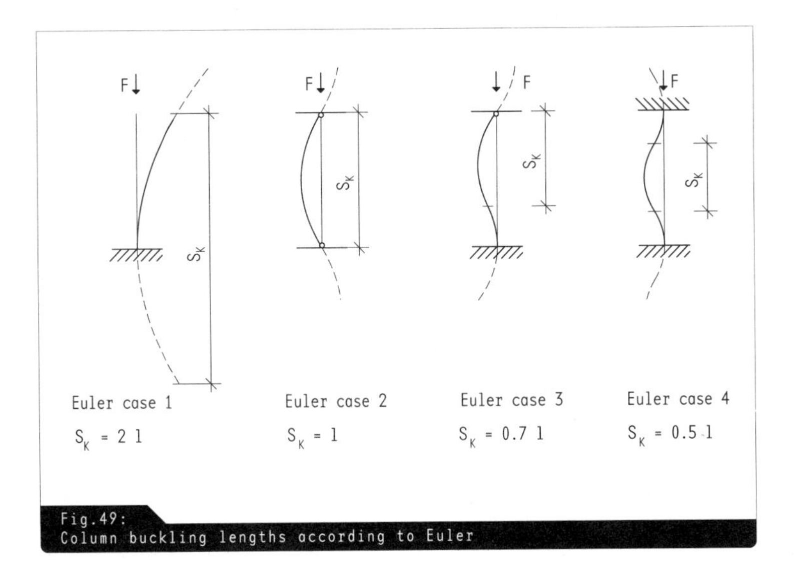

Fig.49:
Column buckling lengths according to Euler

column in relation to the deformation curve is known as the effective or buckling length.

Figure 49 shows the four cases with the same column length. Euler case 1 works on the flagpole principle: the deformation curve is very long, which is unfavourable in terms of stability. Euler case 2 relates to a column that is attached by an articulated joint at the top and the bottom. This case is very common, and the deformation curve or buckling length is shorter, which makes the column more stable. In Euler case 3, the column is braced on one side. This bracing stops the column from twisting at that point and thus reduces the length of the sinus curve, i.e. the buckling length. Euler case 4 with bracing at the top and bottom produces the shortest buckling length for the column, and is consequently the most stable variant.

\\Hint:
Euler's buckling behaviour for columns assumes compression- and tensionproof material such as steel or wood. Euler's scheme is not suitable for dimensioning columns in masonry or concrete.

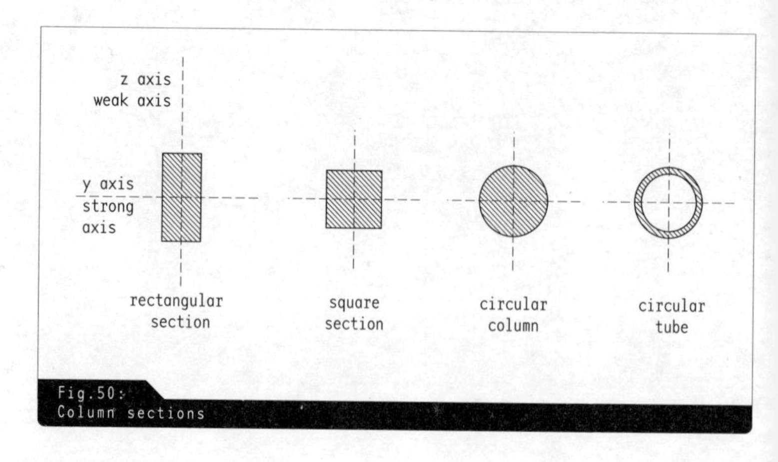

Fig.50:
Column sections

Slenderness

Another major factor for dimensioning a column is its slenderness. It is easy to assume that slenderness implies a ratio of column length to its thickness. But this is not the case: thickness is not part of the equation, but its stability as a ratio of moment of inertia and cross section area are; and then again it is not the length of the column, but Euler's buckling length, as explained above, that is crucial. So the slenderness of a column is the ratio of its buckling length to its bending strength.

It is possible to use these factors involved in calculating columns to make a theoretical statement about the best possible shape for them: columns that are loaded only vertically can buckle in any direction. However, they will actually buckle in the direction in which they have the lowest bending strength. So columns should be equally stable in every direction, as would be the case for a square or – even better – a circular column.

In addition, bending strength in relation to the moment of inertia makes it possible to draw further conclusions about the ideal cross section. With respect to the distribution of tensions in a buckling column it is clear that the areas some distance away from the plane of zero tension or the centre are the most effective. In tubes, the less effective central areas are omitted. Here the material is placed as far away as possible from the central point. This suggests that a tube, and ideally a round tube, is the best possible shape for a column. This deduction is highly theoretical, and is intended only to explain the loading on a column, as ultimately a lot of other factors influence their structure, and all these have to be taken into consideration. › **Figs 50 and 51**

Fig. 51: Columns

CABLE

Cables do not obey any of the rules explained in previous chapters. If a cable is part of a loadbearing structure, it sags according to the load suspended on it or its own weight, and changes its form with every change of load. It cannot resist bending moments, and always takes up the form in which no moment occurs anywhere. This form corresponds precisely with the bending moment of a girder, rather than a cable.

Funicular line

So the "funicular line" corresponds with the cable's moment curvature. › Fig. 52

A second important difference from the loadbearing structures discussed previously is the fact that cable loadbearing structures always have horizontal reactions at the support as well. Cables dissipate all loads as normal forces, i.e. the funicular force, and correspondingly the horizontal reaction force, follow the direction of the cable at the support exactly. Only if the cable were hanging vertically would the reaction at support be vertical alone. › Fig. 13, p. 20 In Figure 53, comparing the two cables shows that the vertical proportion of the funicular force corresponding to the magnitude of the loads remains the same, while the horizontal proportion changes with the angle of the cable, i.e. with its slack.

Sag

Funicular force

Something we have all experienced with a tight or slack washing line is an important factor for all cable loadbearing structures: a low sag level means a high funicular force, and a high sag level a low funicular force.

So why are they not used much more often? Cable-stayed structures have their vagaries in practical applications. The magnitude of the deformations they admit causes great difficulties in built structures. Uncontrolled movements – flapping in the wind, for example – have to be completely suppressed in order to avoid large dynamic stresses. Cable-stayed

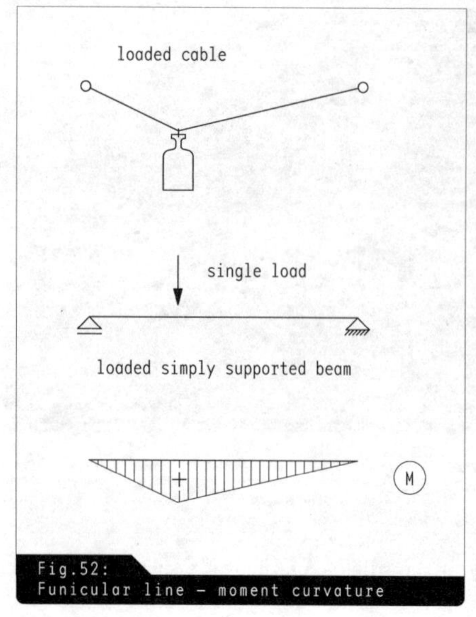

Fig.52:
Funicular line – moment curvature

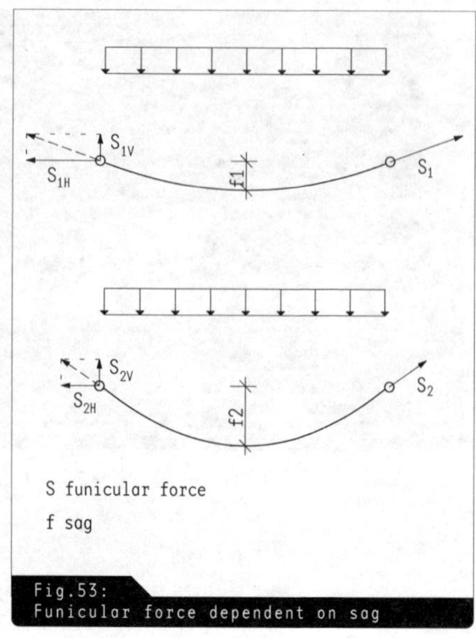

Fig.53:
Funicular force dependent on sag

structures therefore have to be stable in form in every case, which few methods can achieve. One possibility is to weight the cable structure so that possible changes of load or wind loads remain low in comparison with the dead weight of the structure. This solution is available for suspended roofs, for example.

The disadvantage here is that the additional loads, usually applied in the form of prefabricated concrete parts, actually mean losing the advantages of the cable structure, and also that correspondingly large funicular forces have to be dissipated.

\\Hint:
The cables in cable-stayed structures are made of high-strength steel. Many thin steel wires, with a diameter that varies according to cable type, are twisted around each other to form strands. These strands are then made into cables.

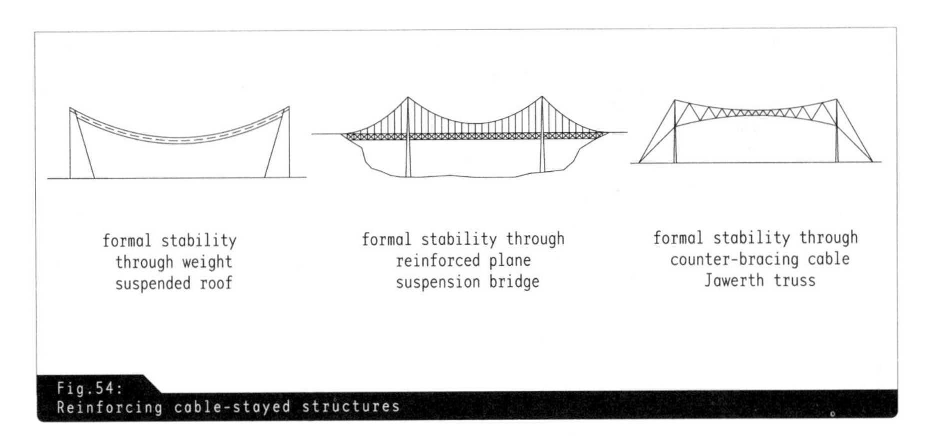

Fig.54:
Reinforcing cable-stayed structures

Another solution is reinforcement with flexurally rigid structural elements. In suspension bridges, for example, the suspended carriageway is so flexurally rigid that it reinforces the whole bridge.

A cable-stayed structure can also be reinforced by counterbracing with additional cables. This can be done in a variety of ways. Two-dimensional beams can be manufactured, e.g. the Jawerth truss, which is reminiscent of lattice structures, but actually has nothing in common with them. All the cables in systems of this kind are so highly pretensioned that no cable slackens, even under the highest possible load. This means that the structure remains stable in form and capable of loadbearing. › Fig. 54 In two-dimensional structural elements made up of cablenets, rigidity in the system is achieved by prestressing areas that curve in opposite directions to each other. › Chapter Plate structures

ARCH

If a loaded cable is fixed and turned over, we have a form that dissipates loads as compression forces, and not tensile forces. This is the ideal arch form, since, like a cable, it dissipates load only as normal forces.

Resistance line

This ideal form, which can be established by calculation or by a drawing method, is called a resistance line.

Arch and cable also have other things in common.

Arch height

The arch also dissipates vertical and horizontal forces in both supports, and as with the cable, the height of the arch, measured from floor or base level to apex, is linked with the magnitude of the horizontal forces: the shallower the arch, the greater the proportion of horizontal forces working as compression forces, known as arch thrust. › Fig. 56

Fig.55:
Arcuated loadbearing systems

The crucial difference between cable and arch lies in the fact that as solid arch, unlike a cable, cannot follow a change of load by changing its shape. A resistance line as an exact arch form applies only to an individual load position. If the load changes, the resistance line changes as well. This means that both normal force and bending moments are created in an arch. There are various ways in which arch structures can deal with these problems.

Masonry arches usually have a very large dead weight. Because the working load is small in comparison with the dead load, there are few consequences for the resistance line if it changes. The arch remains stable. Arches can also be reinforced with additional structural elements. For example, if masonry is raised round an arch in the form of a wall, it prevents the arch from deforming or losing its loadbearing capacity. It is also possible to make arches of rigid materials such as laminated timber or steel. Here, the static height of the arch support must be large enough to be able to absorb the moments as well as the normal forces. › Fig. 57

\\Tip:
Genuine arcuated loadbearing systems should not be confused with arcuated bending beams. An arch whose horizontal forces are not absorbed by both supports can dissipate its forces only by bending.

\\Hint:
Arcuated loadbearing systems are derived from masonry construction. Since masonry can absorb only compression forces, all the apertures have to be spanned by arches. Old masonry structures present an opportunity to study many sophisticated arcuated loadbearing systems and the skilful treatment of arch thrust.

Further information about masonry arches can be found in *Basics Masonry Construction* by Nils Kummer, Birkhäuser Verlag, Basel 2007.

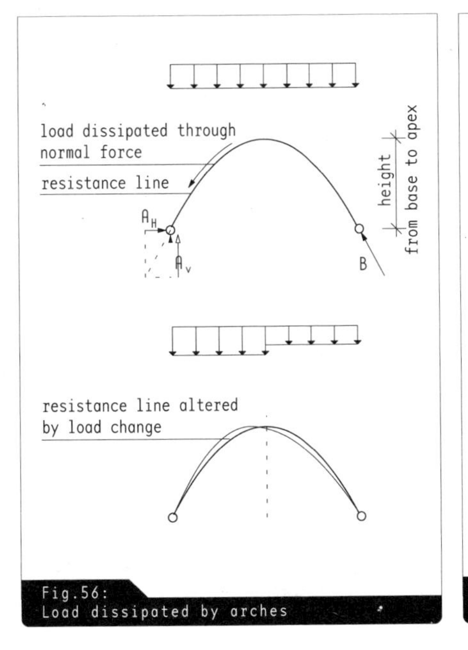

Fig.56:
Load dissipated by arches

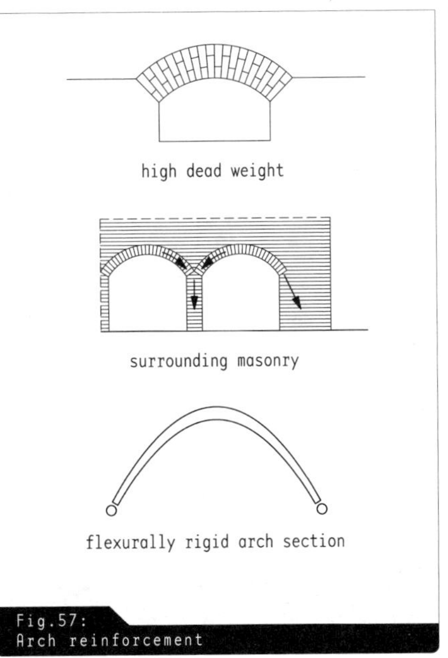

Fig.57:
Arch reinforcement

In arches, we distinguish between three different statical systems: two-articulated arches, three-articulated arches, and arches without articulation.

Two-articulated arch

A two-articulated arch has articulated supports. They absorb horizontal and vertical forces, but no moments. The question of what would happen if a support were lowered shows that this is a statically undetermined system.

Three-articulated arch

Adding one more articulation, usually at the apex of the arch, turns this statically undetermined into a statically determined system. This makes hardly any difference to the loadbearing properties, but the advantage in terms of construction engineering is that an arch is easier to transport in two parts. The articulation is created by the fact that the two parts of the arch are then simply leaned against each other at the apex and screwed together.

Arch without articulation (braced arch)

Bracing the supports makes the arch more rigid because the braces prevent any distortion caused by bending moments. The effect can be compared with columns supported as in Euler cases 2 or 4, and the bracing makes the structure more rigid here as well. Braced arches are statically undetermined. They are very rare because effective bracing requires a very elaborate construction. › Fig. 58

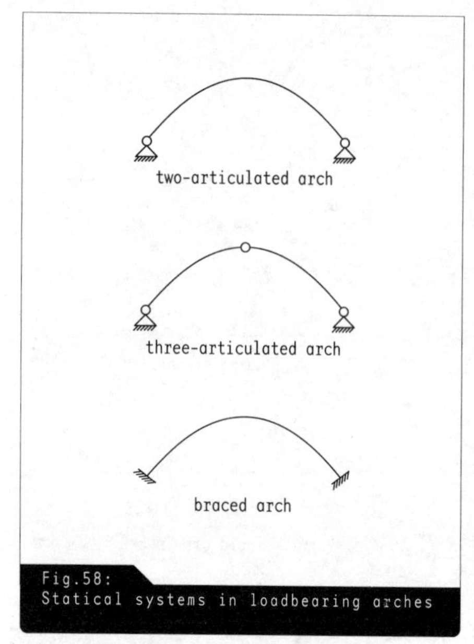

Fig.58:
Statical systems in loadbearing arches

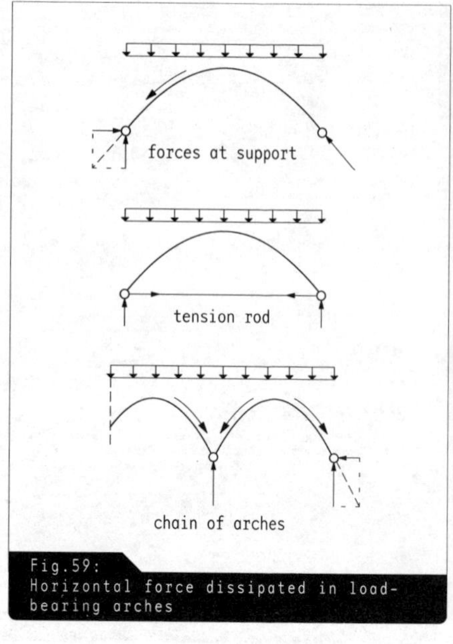

Fig.59:
Horizontal force dissipated in loadbearing arches

Arch thrust

There are various ways of handling the horizontal forces produced. Either the supports can be constructed so that they make it possible to dissipate the arch thrust, or a tension rod can be inserted between the supports to balance the horizontal forces in one with those in the other. If several arches are built adjacent to each other, the horizontal forces acting on the connected supports cancel each other out, so that only vertical forces have to be dissipated. › Fig. 59

FRAME

A simple loadbearing system consists of two columns with a beam or truss above them. But this system is not stable until the columns have articulated support at the top and bottom. Stability can be achieved by connecting the horizontal beam to the columns in a way that is flexurally rigid. This produces an efficient system, a frame.

Rail

In a frame, the horizontal members are called rails and the supports posts.

Post

When the rails and posts are joined rigidly, they behave as though the beam is running "round the corner". So if the rail bends under a load, it also transfers the bending force into the posts. These would deflect outwards if they were not supported. The supports thus resist deformation and the

Fig.60:
Corners of steel frames

stresses are addressed by the structure as a whole. The posts also limit sagging in the rails. So each rail does not function like a simply supported beam, but is partially restrained.

This is also clear from the moment gradient. A characteristic feature of a frame is the moment at support created by the restraining effect of the posts in the corners of the frame. This reduces the moment of span of the frame rail. The advantage for the loadbearing capacity here is that of a continuous, as opposed to a simple, support. The moment at support reduces the moment of span, which means that the dimensions of the beam can be smaller. › **Fig. 61**

It also becomes clear that the corners of the frame are subject to a high load by the moment at support. They have to be constructed carefully, in order to have the required flexural strength. So that the simplest possible structural elements can be prefabricated, it makes sense to manufacture the rails and posts separately and join them together only at the building site. But this exacerbates the problem with the flexurally rigid corners, which nevertheless create another essential advantage for the system. We said at the outset that it is only the flexurally rigid corners that make the frame into a stable system. They reinforce it longitudinally, which is important for skeleton structures. A frame in a statical system has a similar function to a complete shear wall, and can be used to reinforce built structures. › **Fig. 61 and Chapter Reinforcement**

Two-articulated frame

The frames in Figure 59 are shown with two articulated supports. They are called two-articulated frames and, like two-articulated arches, are statically undetermined systems.

Three-articulated frame

A further articulation can be added to frames as well as to arches, to make the system statically determined. This makes very little difference to the loadbearing capacity, but the construction can benefit from it under certain circumstances, especially as this third articulation can be placed

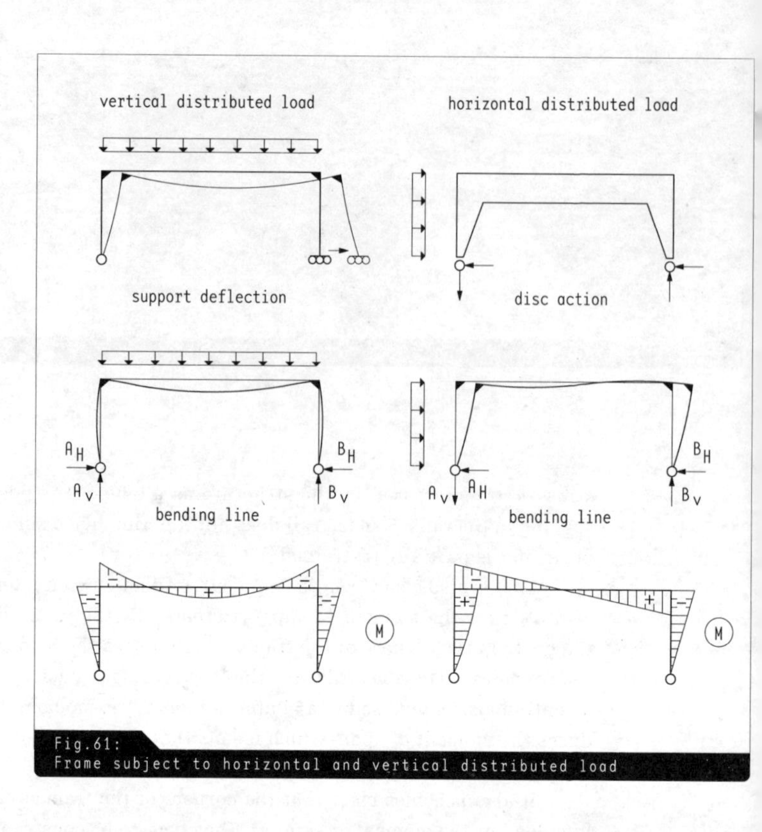

Fig.61:
Frame subject to horizontal and vertical distributed load

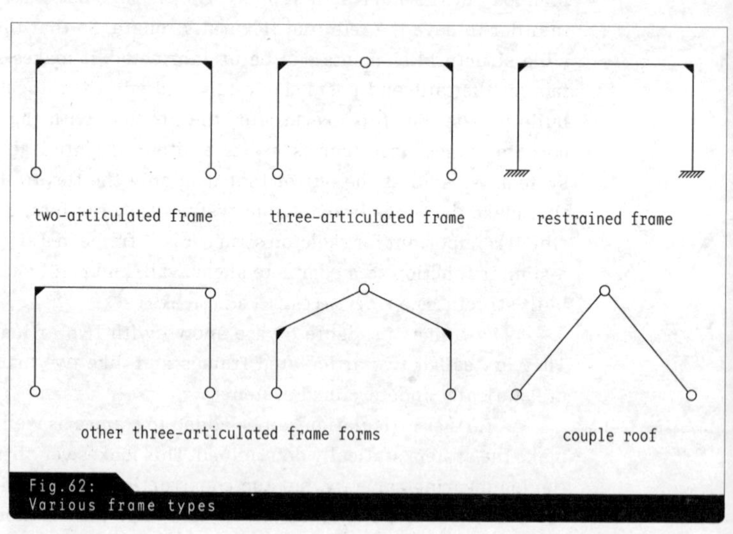

Fig.62:
Various frame types

in a variety of ways. It can be in the middle, in the ridge or even in a corner of the frame. Because the bending moments are zero at the articulation point, the construction can be more filigree here than in the areas with greater bending moments.

Restrained frame

The rigidity of frames can be further increased by bracing the posts into the supports. They are then called restrained frames, but are seldom used, because restraining the posts is a very elaborate process. › Fig. 62

LOADBEARING STRUCTURES

Buildings are complex three-dimensional structures, and at first their loadbearing systems seem immensely elaborate and difficult to analyse. But fundamentally all construction types are derived from two principles: solid construction and skeleton construction. These two principles have been applied since the earliest days of building, and all the techniques so far invented follow them; the same rules apply to ancient clay huts or pile dwellings as to the modern building industry's complex systems. Figure 63 shows a specimen ground plan as a solid structure, a skeleton structure and in some hybrid constructions.

SOLID CONSTRUCTION

Disc

Solid structures are made up of flat elements that dissipate vertical and horizontal loads. Wall-like discs can be loaded vertically as well as horizontally in their longitudinal direction. But conversely, they have barely any transverse loadbearing capacity, i.e. via their surface. › Fig. 64 Discs or walls can fail in a variety of ways; they can buckle or fall over. When building using solid techniques they are protected from this through reinforcement by other walls, placed at certain intervals adjacent to or intersecting them. The walls support each other mutually and this makes a sold structure stable.

Modular construction method

A structure of this type is also called modular. We distinguish between loadbearing, reinforcing and non-loadbearing walls. Non-loadbearing walls can be removed without affecting structural stability. Reinforcing walls are also deemed to be loadbearing in standards. As a rule, loadbearing walls are thicker, which enables them to dissipate the ceiling loads.

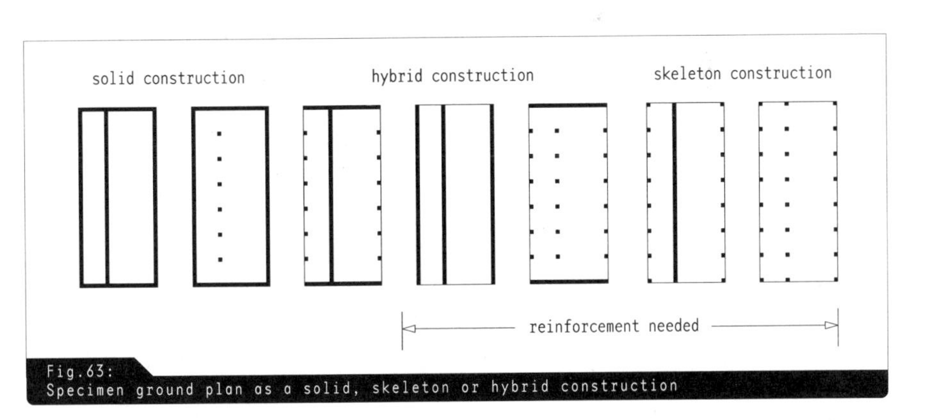

Fig.63:
Specimen ground plan as a solid, skeleton or hybrid construction

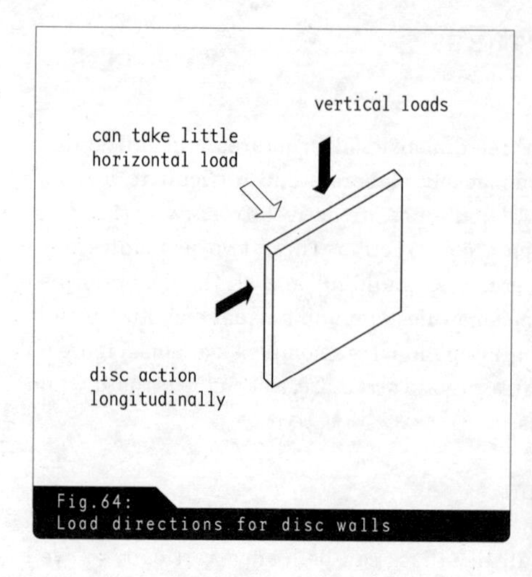

Fig.64:
Load directions for disc walls

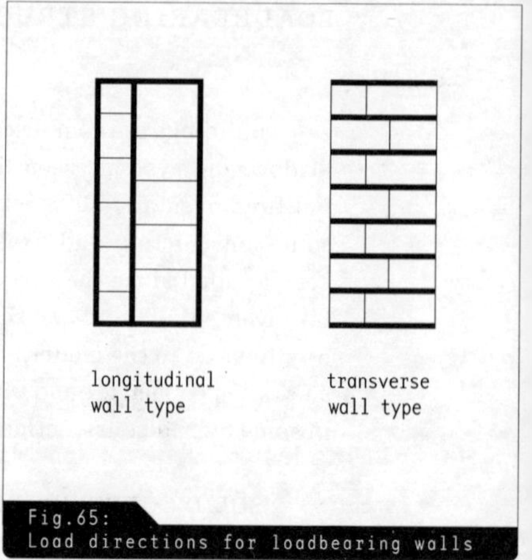

Fig.65:
Load directions for loadbearing walls

Solid structures are divided into longitudinal and transverse wall types.

Longitudinal wall type

If one or two loadbearing central walls run parallel to the long sides of the building, this is known as a longitudinal wall type; most simple, urban homes are built on this principle.

Transverse wall type

The transverse wall type, also known as crosswall construction, is suitable for buildings such as hotels and terraced houses, where small rooms are the principal requirement. It is possible to make a distinction between these types when using floors with timber beams or prefabricated concrete parts with uniaxially directed stresses. When using concrete floors that dissipate their loads in several directions, longitudinal and transverse walls are usually loadbearing. › Fig. 65

\\Hint:
The terms solid or massive construction and skeleton construction do not mean the same thing for architects as they do for structural engineers. The above explanation is couched in architectural language, which is based on geometry and structure. For structural engineers, solid construction is a subject in its own right dealing with masonry and reinforced concrete. So structural engineers tend to link the term solid construction with the material.

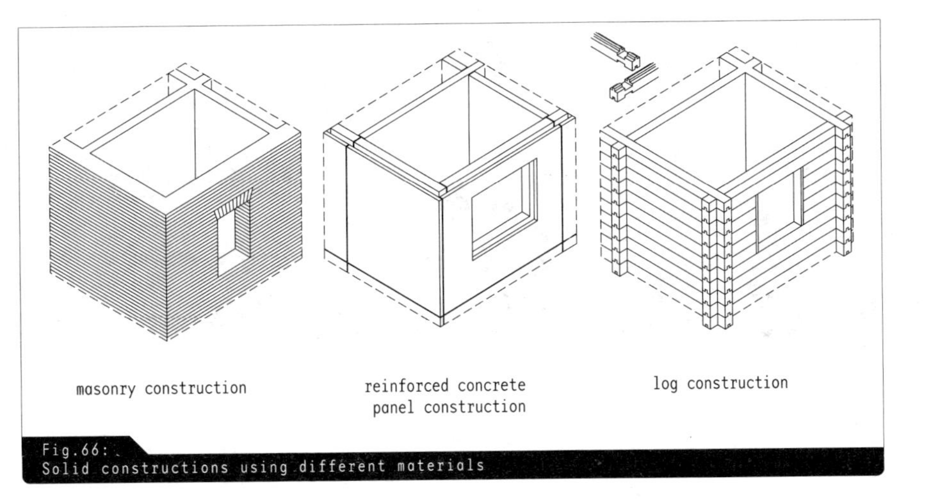

Fig.66:
Solid constructions using different materials

Masonry

The original solid construction is the masonry building. Masonry walls cannot absorb tensile stresses, and have to be reinforced appropriately to their height, length and thickness. Tensile strains are best avoided by clear load dissipation without protrusions, shoring or wide apertures.

Concrete

The chapter on slabs explained that reinforced concrete can also absorb tension forces. This means that concrete walls are considerably more stable than masonry walls, and that solid structures in concrete can be designed with a much greater degree of freedom in relation to room sizes, spans, apertures and structural complexity. They can be cast in situ or constructed from prefabricated parts, which are made up either of small slabs or of wall elements the size of the room, known as large panels.

Panel construction method

Construction using large panels is the popular industrialized building method, and is usually known as large-panel or slab construction. The components are fixed together with steel structural elements and concrete to create a continuous, monolithic structure.

Timber

Although timber construction usually employs the skeleton methods, there are some structures that are better classified as solid construction.

Log construction method

The first is log construction, where timber sections are piled horizontally to construct walls. Walls of this kind are stabilized by the halving joints used for the timbers at the corners of rooms or the whole building. The timber industry has progressed in recent years to the extent that for a few years now there have been panel materials on the market that make panel construction possible. Some of these are panels glued together from planks, like laminated timber, and some are plywood panels

Fig.67:
Skeleton construction

made from boards layered crosswise. These panel materials make construction methods viable that are very different from traditional timber construction methods. These methods are still in development at the time of writing.

SKELETON CONSTRUCTION

Skeleton constructions are made up of bar-shaped elements forming a structure like scaffolding. Panel and wall elements are then added to this structure. The loadbearing structure and the elements that create the interior spaces are, in principle, two separate systems. › Fig. 68

Fundamentally, skeleton structures are made up of three kinds of structural elements: the columns, the floor beams including the floor structure, and the reinforcement structures that absorb horizontal forces. These structural elements are fitted appropriately to the material at nodal points, almost always with articulated joints. Joints are articulated if they are not rigid enough to act as a restraint. They do not have to take the form of a hinge or similar. In principle, any material that is both compression- and tension-resistant can be used for skeleton structures, for example, timber, steel or concrete. Each of these has its own construction methods, with a particular set of problems arising from the material and the methods used in jointing it.

Concrete

Probably the commonest material for skeleton structures is reinforced concrete, with both in-situ cast concrete and reinforced elements as viable possibilities. A solid reinforced concrete floor slab is normally used for in-situ concrete structures, and then only reinforcements and columns are needed. This simplicity also explains the flexibility and economic viability of the system. › Fig. 69a But floors, as point-supported flat structures, admit only limited spans. All the forces from the floor slabs have to be transferred into the columns, which means that the points of transition

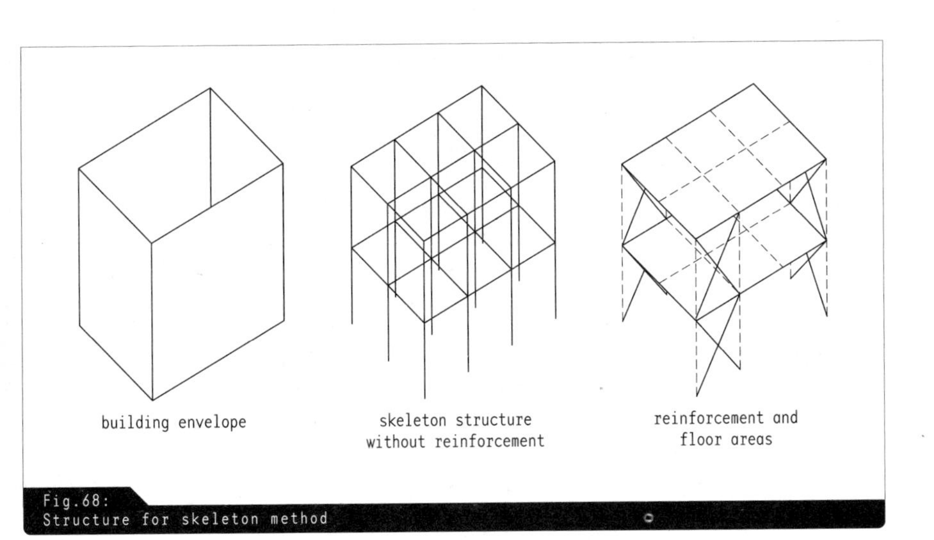

Fig.68:
Structure for skeleton method

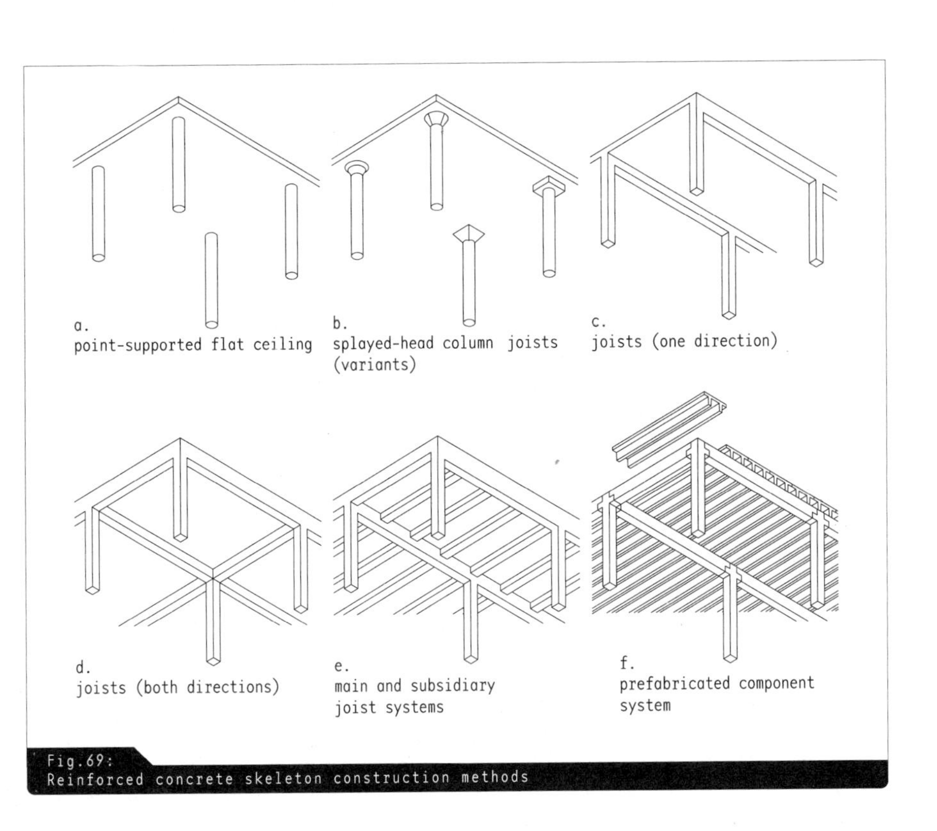

Fig.69:
Reinforced concrete skeleton construction methods

from column to floor are very heavily loaded. There is a risk of the column punching through the floor slab.

Splayed-head columns

To avoid this, the edge can be reinforced in a different way. One method is to use "splayed-head columns". › Fig. 69b

Joists

If the spans are too great for this system, joists are used. These run from column to column like beams, and support the floor slabs on a linear pattern. Joists can be arranged in a number of different ways. According to the span, they are planned to run in one direction, in both directions, or as a system of main and subsidiary joists. › Fig. 69c, d, e

Skeleton structures can also be erected using prefabricated elements. There are various prefabricated systems containing components for ceilings, joists, columns and foundations. Transport to the building site is a crucial factor in terms of size. The structural elements should ideally not exceed the stipulated dimensions for a lorry-load in order to be financially viable. All that usually happens at the building site is that the parts are placed on top of one another and fixed securely into position. This means that the joints are articulated in principle.

Pi plates

Pi plates are often used as prefabricated floor components instead of flat floors. They are narrow plates with two ribs, and can be fitted together to make floor slabs. They work on the T-beam principle and thus make large spans possible. Floor slabs of this kind are laid on concrete beams that are supported on the column brackets in their turn. › Fig. 69f and Chapter Slab

Steel

Steel structures are almost always skeleton structures. They are usually built up of standard steel construction sections, called "rolled sections", in different profile series. › Fig. 70 The size needed is established by statical calculation.

Rolled sections

Rolled sections are produced to a height of up to 60 cm. If higher units are needed, they have to be welded from sheet metal; in steel construction, elements up to several centimetres thick are still known as sheets.

Corrugated sheets

Corrugated sheets are generally used if large areas have to be covered. They acquire their loadbearing capacity from their trapezoid folds,

\\Hint:
The nature of the floor structure greatly affects the clear height of each storey, so they should be included in the structural considerations as early as when the first sections are drawn. The greater the width of the floor spans, the greater their structural height.

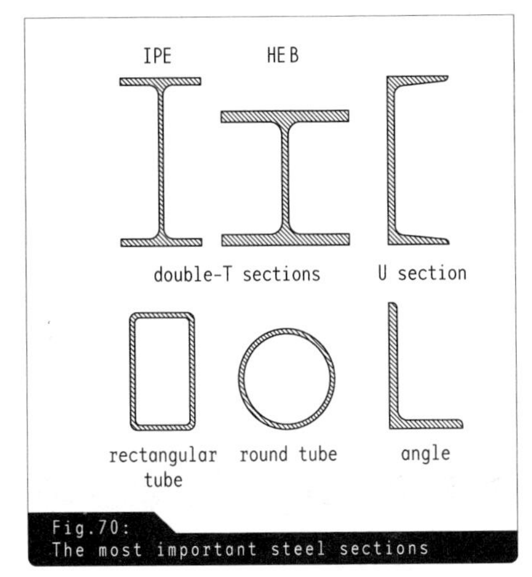

Fig.70:
The most important steel sections

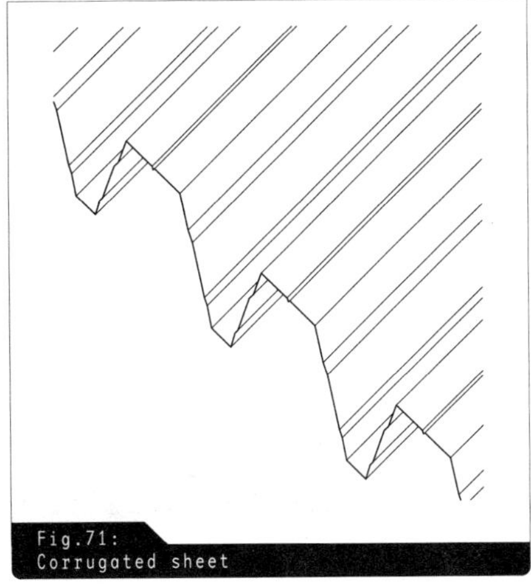
Fig.71:
Corrugated sheet

and are able to function over large spans and serve as floor or roof structures. › Fig. 71

As a rule, steel structural elements are made in a steel construction workshop in transportable sizes and are then assembled on the building site. In the workshop, welding is the simplest and best method for manufacturing steel structural elements, but as welding is difficult on the building site, screws should be used for assembly connections.

It is possible to create flexurally rigid corners with an acceptable degree of effort and expense in steel construction. This means that columns and beams can be fitted together to form loadbearing frameworks to exploit their reinforcing capacities. › Chapter Frame To handle the very strong forces at the frame corners, the connections have to be much more powerful at these points. For example, for double-T sections, both flanges on a

\\Tip:
All the current construction reference works give steel construction section tables with precise dimensions and statical values. Generally speaking, sections from these series should be used in construction, as they are obtainable for any steel construction firm and economical in use.

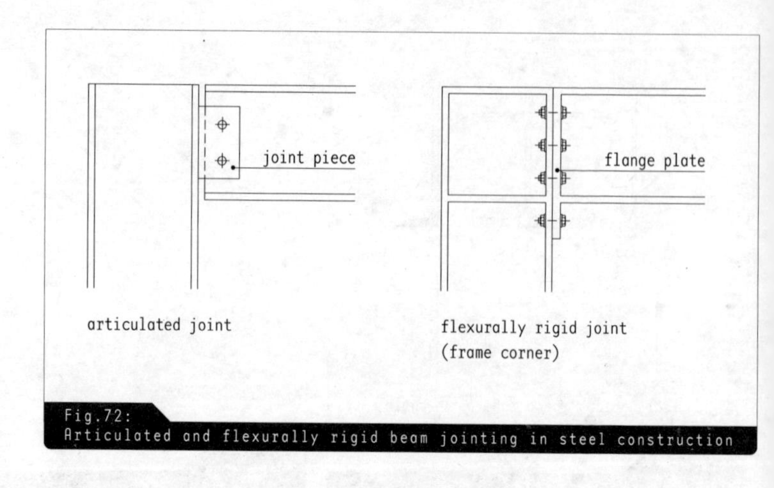

Fig.72:
Articulated and flexurally rigid beam jointing in steel construction

horizontal member have to be fastened to the post with a flange plate, and the connecting screws should be as far away from each other as possible. Conversely, for an articulated joint, the rib can be screwed with a simple sheet metal joint piece. › Fig. 72

Fire protection

Although it seems remarkable at first, steel structures are at greater risk in fire than timber structures. Steel softens when heated to high temperatures, and quite quickly loses its entire loadbearing capacity. Steel must therefore always be protected from fire in high-rise buildings, for example by cladding loadbearing members with plaster or with a foaming paint.

Composite constructions

The rate at which the steel heats up in a fire can also be lowered by installing it in combination with concrete. For example, tubular steel profiles can be set in concrete in these composite constructions, or double-T sections filled with concrete. As well as slowing down the heating process, the concrete will ensure a certain residual loadbearing capacity in the event of fire. › Fig. 73

Timber

Timber is the earliest timber construction material. Various cultures have some very old, but very sophisticated timber construction techniques. This is a complex matter, because there are a number of construction methods and an infinite number of variants and mixtures between them. Here are the most important categories:

›

Traditional timber-frame construction

Traditional timber-frame construction is a pure form of skeleton construction filled with clay or brick. As a craft construction method, it is characterized by joints using skilfully created forms, without any metal connecting devices. Timber-frame buildings are seldom constructed like this any more, but they are often found in the field of heritage.

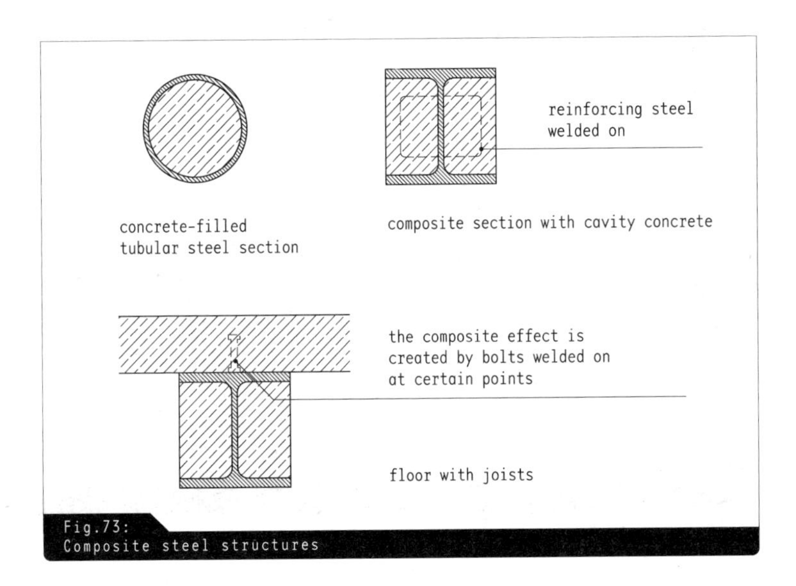

Fig.73:
Composite steel structures

Balloon and platform frame construction

American timber construction using the balloon and platform frame methods differs from traditional timber-frame construction in that it uses thin log- or plank-like timbers that would have no loadbearing capacity in their own right and would buckle if the timber cladding that forms the wall surface did not hold them in position. They work like ribs; they are stable with the cladding. The method is therefore sometimes called rib construction. Nailing is the principal jointing device. Constructions of this type are highly economical and flexible.

Engineering timber skeleton constructions

Modern engineering timber skeleton constructions have an ideal loadbearing system from a statical point of view, and can be constructed differently according to use. Materials such as laminated timber or various sheet products are used.

Modern timber-frame construction

Prefabrication is becoming increasingly accepted in timber skeleton construction. Here, wall and floor elements, in dimensions that can

\\Hint:
More information on timber construction can be found in *Basics Timber Construction* by Ludwig Steiger, Birkhäuser Publishers, Basel 2007.

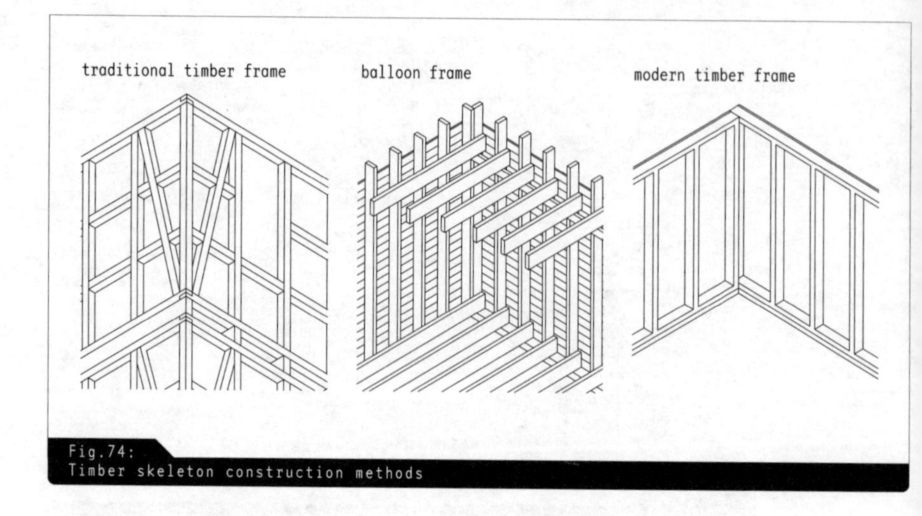

Fig.74:
Timber skeleton construction methods

be transported on lorries, are preferred as component sizes. Timber-frame construction seems best suited to this. Prefabricated parts are manufactured consisting of derived timber sheets, onto which loadbearing sections are screwed. The parts can also be supplied with built-in insulation, cladding, windows or doors, and then fitted together. Similarly to the American methods, the frame sections work with the laminated wood areas to form a loadbearing framework. › Fig. 74

REINFORCEMENT

When planning skeleton constructions the key aim is usually to dissipate dead weight and vertical working loads, for which floors and columns are constructed. But attention must be paid to horizontal loads as well. The most important horizontal load is wind load, which can act on the building in any direction. Because the component joints are generally articulated, skeleton constructions have almost nothing to resist horizontal loads. They therefore need effective reinforcement, i.e. a construction that can transfer the horizontal loads from the façades into the foundations.

Reinforcing constructions function as a disc. They can accept horizontal forces longitudinally and dissipate them downwards. In tall buildings they work as vertical loadbearing members that can dissipate wind loads into the foundations from all floors.

Disc action

A disc can be solid, usually made of masonry or concrete. Disc action can also be created by a diagonal brace in one compartment of the skel-

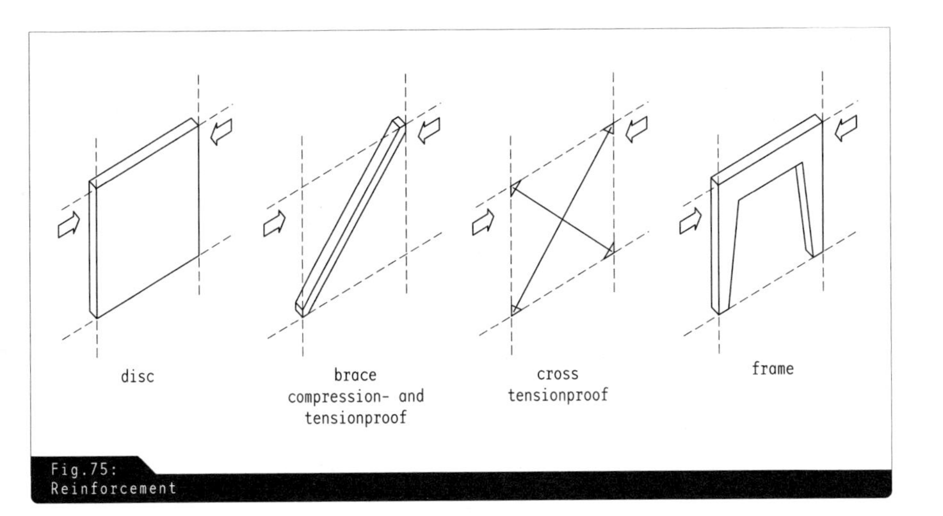

Fig.75:
Reinforcement

eton structure. This brace reacts to compression for loads in one direction, and to tension for loads in the other. The same effect is achieved from two tensionproof crossing diagonals. › Fig. 75 The reinforcing action of frame systems was also pointed out in the chapter Frames.

Skeleton constructions have to be reinforced transversely and longitudinally. Reinforcement in each direction is not sufficient because, considering the ground plan, two disc elements always intersect at one point. This intersection would then be the point around which the loadbearing structure could twist and would collapse. To prevent this, we need a new plane of reinforcement, which can be positioned as wished, but it must not intersect with the other two at the same point. › Fig. 76a

Reinforcing structures can be arranged on the ground plan in different ways. But they should be placed near the centre, because otherwise the long section of the building would acquire a long lever arm around this reinforcement, thus creating powerful forces that would place an unnecessary strain on it.

Floor disc

If a skeleton construction is loaded horizontally, all the forces from one direction must be transferred into the wall disc provided for the purpose. This needs a rigid floor disc, as assumed in Figure 76. A floor can also consist of joists with a covering on top of them. A floor of this kind is not a disc, because the joists can shift in relation to each other. Not all the horizontal forces can be transferred into the reinforcing structure, but intermediate floors can easily be made into rigid discs by adding braces or cross-braces. › Fig. 77

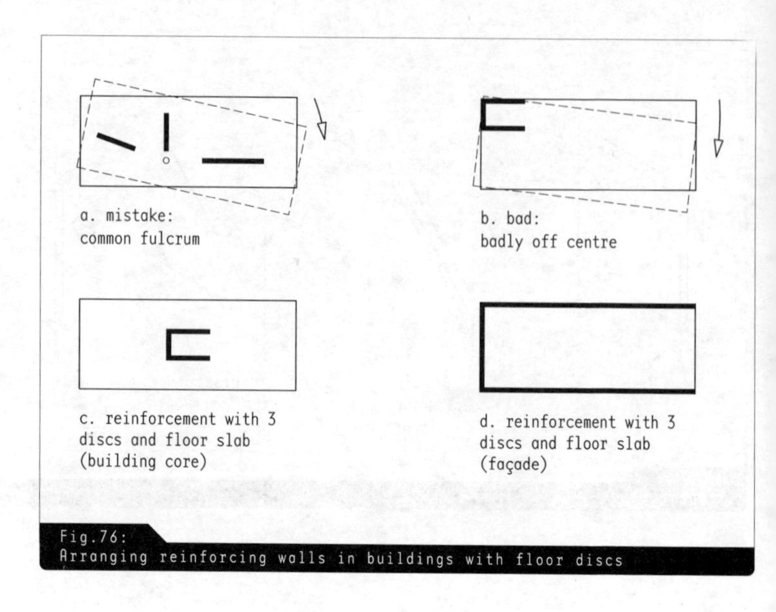

Fig.76:
Arranging reinforcing walls in buildings with floor discs

Building core

In high-rise buildings, building cores containing fire escapes and lift shafts are often used as reinforcing structures. They consist of mostly closed walls and run from the roof to the foundations, and can act as vertical loadbearing members. In high-rise building, it can be more problematic to dissipate the horizontal loads than the vertical ones, because wind speeds increase with the height of the building and the effect of the wind loads is much greater. Although the building core provides reinforcement in most high-rise buildings, one possibility is to make the whole façade of the building function into a vertical trussed girder, thus working with the maximum girder dimensions, i.e. the whole width of the building.

For architects, the key question is whether their design will be adequately reinforced or not, or put it another way, whether it is stable or not. In addition, a distinction is made between less rigid or more rigid loadbearing systems. This depends on how generously or how centrally the reinforcements are arranged. The different reinforcement methods do not have identical effects, and here too a distinction can be made between more or less rigid ones.

HALLS

The hall concept ultimately means little more than the fact that large spaces are enclosed, as they can be built using both solid and skeleton methods, and take any conceivable form. What they have in common is the

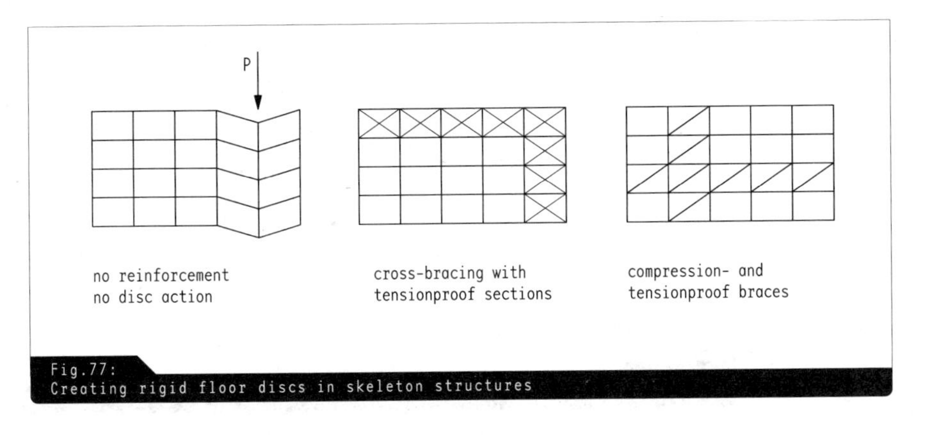

Fig.77:
Creating rigid floor discs in skeleton structures

large span for the roof loadbearing structure. The roof geometry can also be designed in a variety of ways. It can be focused on draining the surface of the roof, on an advantageous shape for the roof girders, or on the way in which the skylights are built into the roof area.

Shed roof

Skylights can be placed longitudinally, or be constructed in the direction of the loadbearing structure and as an integral part of it, as in shed roofs. › **Fig. 78**

Halls thus need a roof loadbearing structure that can handle a large span. It is an advantage here to the make the roof area a lightweight structure, as the dead weight is an additional load on the structure as a whole.

A large number of statical systems are available for hall construction. The most common are described briefly below.

Truss

Long girders resting on columns or walls are also known as roof frames or trusses. Because their support points are articulated, a construction of this kind must be reinforced either by creating a rigid roof disc and the façades, or by bracing the columns. Roof trusses can be made of wood, steel or concrete. › **Fig. 79**

Arch

Arches make appropriate loadbearing structures for large spans, and thus for halls, because the loads are dissipated mainly as normal forces and not as bending forces. A solution must however be found for the great horizontal forces at support. The arches either run to the floor, so that the arch thrust can be transferred into the foundations directly, or they sit on columns or walls, which then have to be reinforced with structures such as buttresses. It is possible to use tie members between the supports to balance the horizontal forces on both sides. Then only the vertical forces have to be transferred into the walls. › **Fig. 80 and Chapter Arch**

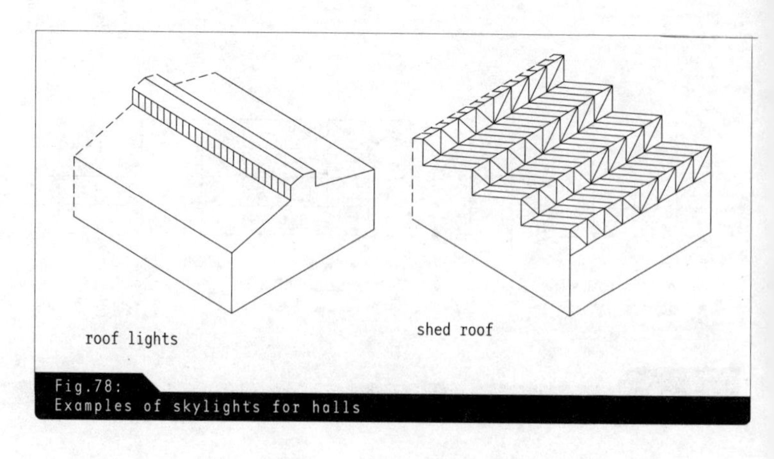

Fig.78:
Examples of skylights for halls

Frame

Frames are well suited to hall construction. They can be used to create all kinds of roof geometries, unlike arch constructions. Asymmetrical forms can also be implemented very well with two- and three-articulated frames. The section dimensions must however always match the moment gradient, which must be established for the particular geometry and load pattern. › Fig. 81 and Chapter Frame

Beam grid

The systems named so far consist of girders, spanning the space in one direction. These are directed systems. But it is also possible to design loadbearing structures that dissipate their loads on all sides. Load dissipation on several sides makes sense primarily for spaces with approximately equal spans in both directions. Here, the girders cross over each other, thus forming a grid. Girder grids of this kind can be made of various different materials. They can run through joists in conjunction with an in-situ cast concrete ceiling, and thus form a monolithic bond. Flexurally rigid connection of each intersection point is more laborious at the assembly stage for steel and timber.

Three-dimensional frameworks

Trussed girders can also be extended to form a three-dimensional loadbearing system. These are then called three-dimensional frameworks, and are defined by the design of the bar and node components. Three-dimensional frameworks are almost always made of prefabricated steel elements. › Fig. 82

Reinforcement

Reinforcement or stiffening for halls obeys the laws explained in the previous chapter. But other points must also be borne in mind. For example, reinforcing just one loadbearing axis is not sufficient for halls above a certain size, because the loads within the loadbearing structure travel a great distance before they are transferred, and so the structure as a whole would not be rigid enough.

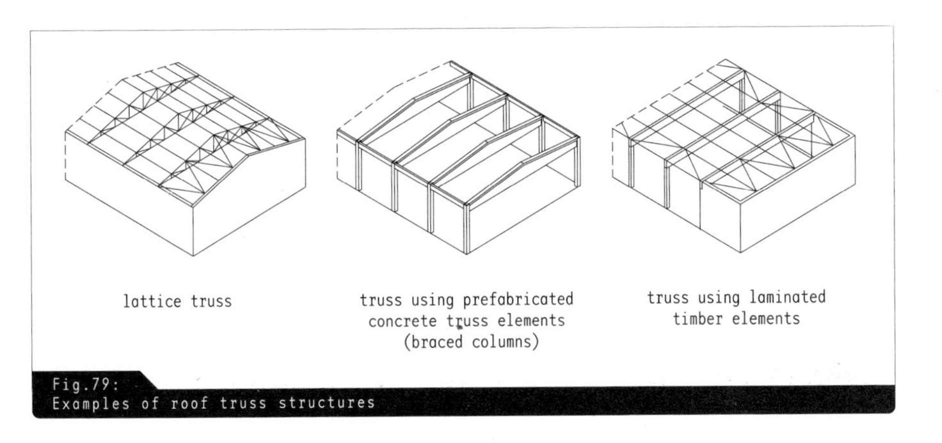

Fig.79:
Examples of roof truss structures

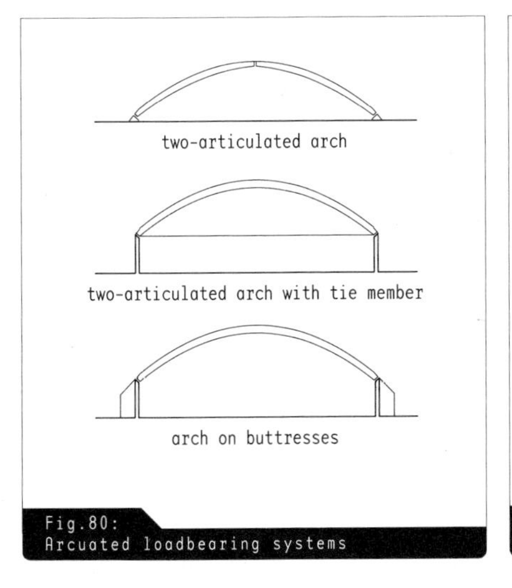

Fig.80:
Arcuated loadbearing systems

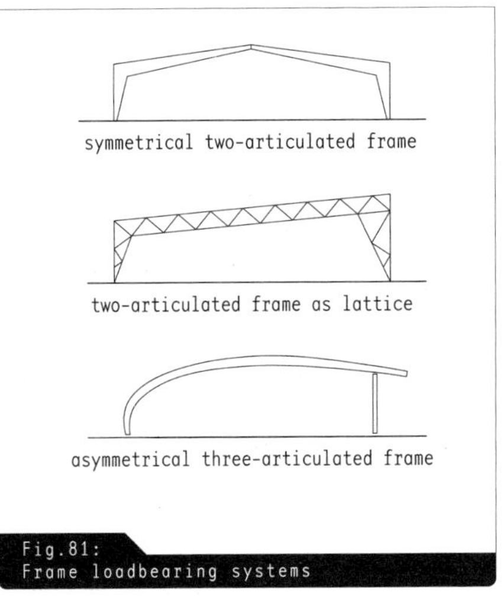

Fig.81:
Frame loadbearing systems

Long girders dimensioned appropriately for their span are at risk of buckling in their transverse direction. › Fig. 44, p. 43 There is a risk of failure through great imposed loads, or through wind load against the gable. To prevent this, a joining construction is usually added at roof level to reinforce the gables and transfer the wind loads to the eaves. Adding members that run transversely to the girders, the purlins, means that the additional trusses are attached to the reinforced areas and thus prevented from buckling. › Fig. 83 and Chapter Reinforcement

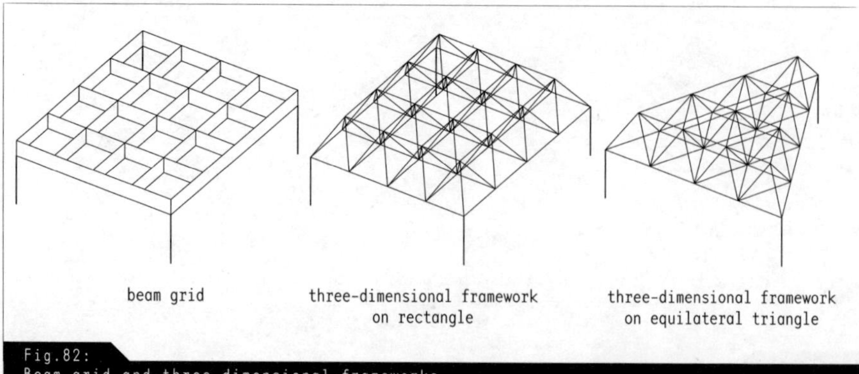

Fig.82:
Beam grid and three-dimensional frameworks

Steel is an ideal material for hall construction. Steel loadbearing structures are light, have a high loadbearing capacity, and can be used to implement all conceivable statical systems economically. One great advantage here is that flexurally rigid joints can also be created easily.

Wood is also a very efficient material for hall construction. Arches, trusses or frames in laminated timber, or lattice trusses made of solid timber sections can be used. Flexurally rigid joints for frame systems can be created with laminated timber.

Concrete halls are always made from prefabricated parts. Their loadbearing systems differ from those in other halls in that the columns are usually clamped into the foundations, but the trusses are articulated at the support points. Reinforced concrete trusses are usually manufactured in adjustable steel formwork, so the system allows little flexibility in choosing the girder geometry.

PLATE STRUCTURES

The chapter Structural elements discussed arches and cables, which dissipate their loads as compression or tension forces, unlike girders subject to bending loads. This load dissipation principle can also be implemented in three dimensions by using plate structures. A large number of different concepts and many variants occur in their design. The most important groups are named below, to give a general idea.

Folded plates/shells

Folded plates are made up of flat surfaces and acquire their loadbearing capacity from the disc action of these areas, while shells are curved loadbearing systems that can differ considerably in their forms. › Fig. 84

Beam-like plate structures

Shells or folded plates can span from support to support as long sections, like beams. Fitting them together then forms a roof. When handling

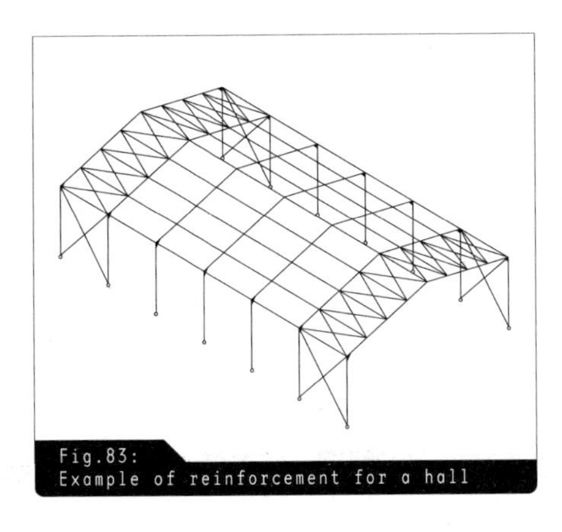

Fig.83:
Example of reinforcement for a hall

large spans, it is important to construct a great statical height with the lowest possible dead weight. Beam-like plate structures are very suitable for this, because of their curved or folded girder sections. Their loadbearing action is most closely allied to that of corrugated sheets. › Fig. 71, p. 65 Beam-like plate structures must be supported at the edges, to prevent them from collapsing as a result of lateral deflection. › Fig. 85

Tension-/compression-loaded plate structures

As with loadbearing systems based on arches and cables, plate structures are distinguished according to their loading type.

Domes and shells

Domes, shells and similar loadbearing structures are compression-loaded in some areas and tension-loaded in others. The more continuously their periphery can be supported, the better they will dissipate loads.

Cablenets and membrane constructions

Conversely, all suspended constructions, such as cablenets and membrane constructions, are only tension-loaded. Concrete structures can be tension-loaded as well. They are supported by flexurally rigid peripheral beams or cables. Peripheral cables of this kind then dissipate the powerful tensile forces through guys that run into foundations with tensionproof anchorage.

Single-/double-curved surfaces

Single-curved surface curve in one direction, but are linear in the other. All curved surfaces that can be made from a flat surface, such as a sheet of paper, are single-curved. They are always sections of cylinders or cones. They can be supported at their ends, like beams, or on their long sides. Unlike beam support, a longitudinally supported, single-curve shell dissipates its loads according to form by the same action as an arch.

Double-curved means that the shells cannot be formed from flat surfaces. Figure 86 shows some examples of this. A double curve makes the surfaces rigid in three dimensions. It ensures that tension-loaded surfaces such as cablenets and membranes will not deform, provided there has been

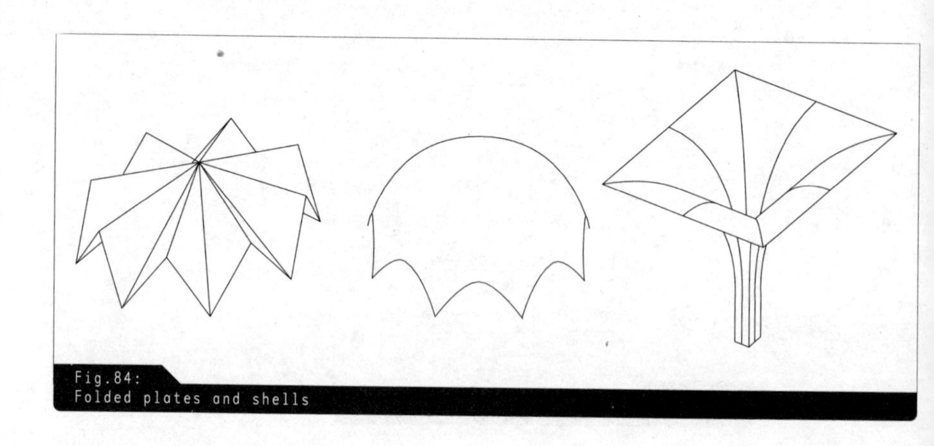

Fig.84:
Folded plates and shells

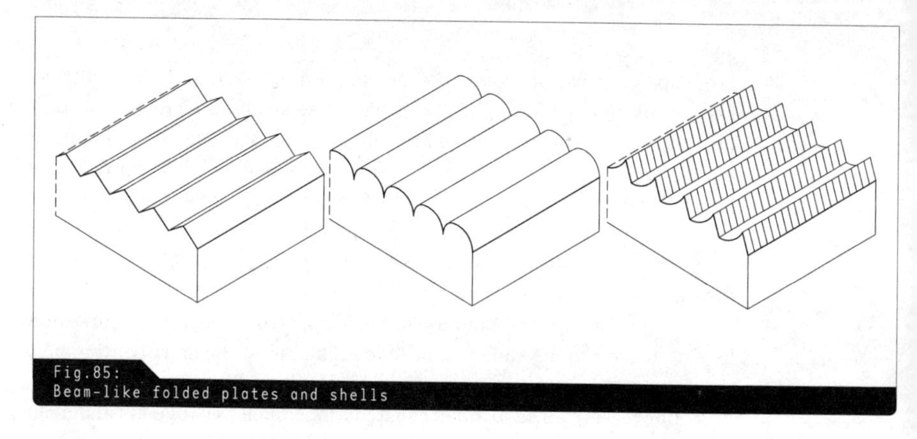

Fig.85:
Beam-like folded plates and shells

adequate pretensioning. Compression-loaded brick or concrete shells thus form loadbearing surfaces even when the material is not very thick.

Single- or counter-directional curved surfaces

Shells or domes are double-curved plate systems operating in one direction. Both curves point in the same direction. Counter-directional curved surfaces are also called saddle surfaces and usually occur in cable-nets or membrane constructions.

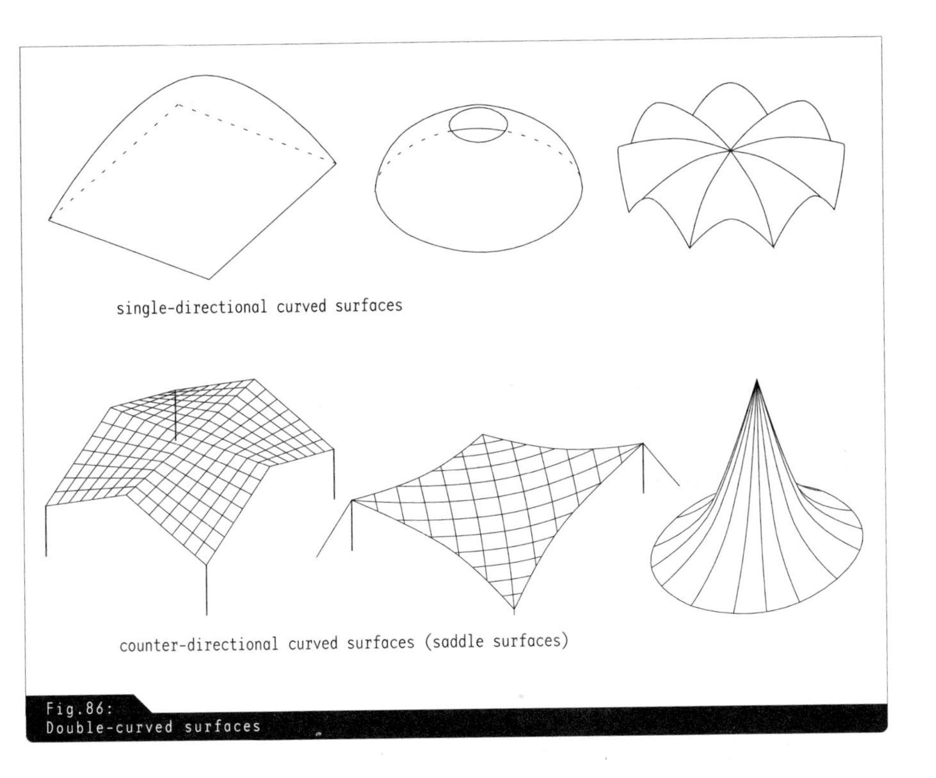

Fig.86:
Double-curved surfaces

FOUNDATIONS

Subsoil

The subsoil is part of the loadbearing structure, as well as the foundations, and like all the other structural elements it must be able to handle the forces it has to accept. Like every other building material, it responds to loads with deformations, which can involve sinking several centimetres. Sinking is thus a normal facet of loadbearing behaviour and is not deleterious.

The subsoil usually has a much lower loadbearing capacity than other building materials. In order to prevent the acceptable stresses being exceeded, the loads to which the building subjects it must be distributed over an adequately large foundation area. Loads spread over a wide area in the subsoil, which means that the stress under the foundation dissipates rapidly with increasing depth under the footing.

Soil type

There are many types of soil, and they respond to loads in different ways. The key factor affecting its properties is the grain size or the grain size mixture. The way the soil responds to fluctuations in humidity is also important. It is therefore essential to collect as much information

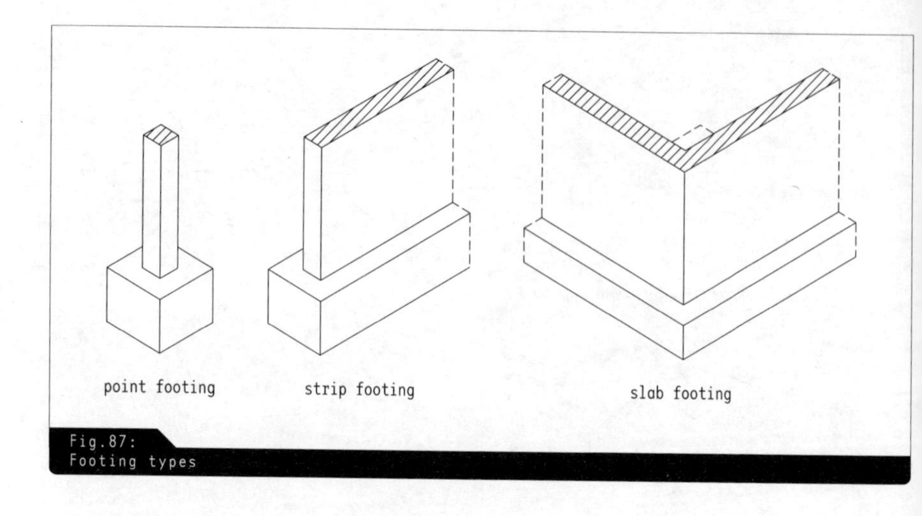

Fig.87:
Footing types

as possible about soil material and soil humidity, and about groundwater levels. A soil report is now a customary feature of smaller construction projects as well.

Footing types

Footings transfer loads into the subsoil. Soil stresses thus depend on the area across which the loads are distributed, i.e. on the size of the footing. A distinction is made between the following footing types:

_ Point footings are usually deployed to absorb the load from individual columns
_ Strip footings dissipate loads from walls, into the ground, for example
_ Slab footings consist of a continuous concrete base that distributes the loads from the walls and columns standing on over the whole area of the building. › Fig. 87

Footings can also be supplied to the building site in prefabricated form, although this makes financial sense mainly for point footings. Figure 88 shows a bucket footing, into which the column fits as if into a bucket. Once the prefabricated column has been adjusted precisely, the joint between the footing and the column is filled with mortar, thus bonding the two prefabricated elements securely.

Deep foundations

If no loadbearing subsoil is to be found in the upper strata, it is possible to dissipate the loads by using deep foundations. To do this, holes are drilled down to the firm stratum and then filled with concrete. These bored piles then act as long columns on which the building stands in the

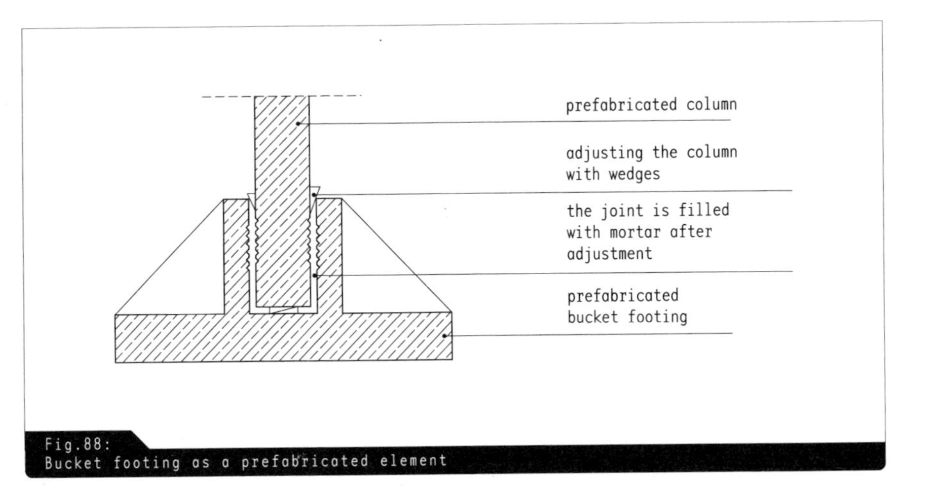

Fig.88:
Bucket footing as a prefabricated element

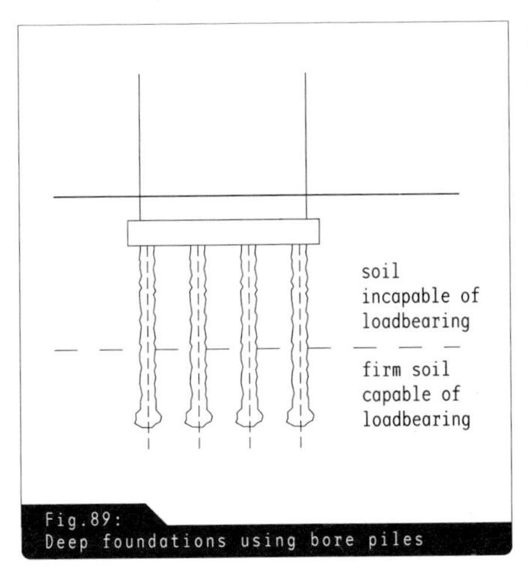

Fig.89:
Deep foundations using bore piles

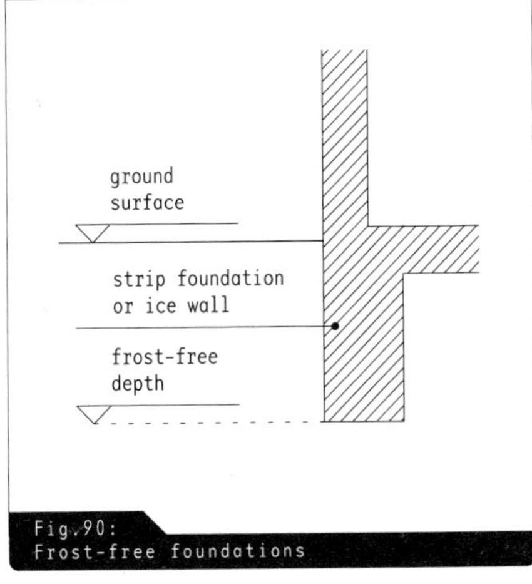

Fig.90:
Frost-free foundations

subsoil. The loads are then largely transferred via the tip of the bored pile, but a firm hold in the subsoil can also be gained via the roughly concreted surface shell of the pile. › Fig. 89

Frost-free foundations

When the ground freezes, it expands because of the increased volume of the ice it contains. This produces perceptible uneven soil deformations. It is therefore necessary to avoid frost action under the footings. The ground only freezes to a certain depth below the surface in winter,

so a continuous strip foundation is laid around the edge of the building, extending into the frost-free area underground. The required depth depends on the climate and can be anything from 80 cm to over a metre.
› Fig. 90

Damaged foundations

Cracks in the completed building usually indicate damaged foundations. Such damage is always the result of irregularities, possibly in the building, possibly in the subsoil.

Changing soil properties inevitably cause problems, because generally speaking each soil type is subject to different degrees of sinking. In terms of the building, problems can arise because parts of the building that impose very different loads share foundations, or from having foundations at different depths, because this will always cause different tensions in the subsoil. It this is recognized at the planning stage, suitable measures can be taken, either to dissipate the loads into the soil evenly, or to avoid possible damage from different sinking rates, for example by means of gaps in the structure.

IN CONCLUSION

Basics Loadbearing Systems is intended to provide an approach to the complex field of loadbearing system theory. The knowledge collected here should enable students to understand structural contexts, to consider the demands made by support and loads when designing, and thus to plan their designs realistically and holistically. Designing the loadbearing structure ultimately helps to sharpen designers' ideas of space and can also further them by working creatively with the possibilities that supporting structures offer. Thus, the quality of the loadbearing structure design is assessed first and foremost by whether it flows from the design idea, or even helps to shape it. This happens primarily in the planning tasks whose function and structure make the loadbearing system the determining element – such as when using large spans. Problems of this kind can usually be solved only by addressing the design of the loadbearing system in its full complexity.

Thus, the basics this volume conveys can be expanded upon as part of the student's own architectural development process by working creatively or even playfully with loadbearing structures, and by interpreting the laws of support structures in terms of individual requirements. To sum up, there are three basic principles to be considered:

1. Loadbearing structural elements should run through all floors to the foundations in a single line.
2. Spans should be kept as small as possible. Large spans demand a great deal of expense and effort. They should be deployed only when large spaces expressly demand them.
3. It is possible to handle large spans without difficulty if sufficient height is allowed to construct them. Even if nothing is known at this stage about the nature of the structure, sufficient statical height should simply be allowed for such spans.

Anyone who would like to go beyond the material introduced in this book and find out more about loadbearing structures will gain a better understanding of how structural engineers work; it is then possible for students to make their own calculations, thus increasing their ability to determine dimensions more precisely, and enabling themselves to design on the basis of the loadbearing structure itself.

APPENDIX

PRE-DIMENSIONING FORMULAE

The formulae below can provide provisional results for dimensioning structural components at the preliminary design stage. They do not furnish conclusive proof of loadbearing capacity.

Floors and ceilings

Concrete floors or ceiling as flat units in multi-storey buildings:

_ Viable at spans up to 6.5 m
_ The formulae apply to simply supported beams
_ Thickness to provide necessary sound insulation at least 16 cm
_ As a point-supported flat ceiling or carried on walls

at a span less than 4.3 m

$$h(m) \approx \frac{l_i(m)}{35} + 0.03 \text{ m}$$

at a span greater than 4.30 m and given limited deflection as a result of light dividing walls on the floor

$$h(m) \approx \frac{l_i^2(m)}{150} + 0.03 \text{ m}$$

Timber beam floor or ceiling:

_ Distance between beams 70–90 cm
_ Width of beams $\approx 0{,}6 \cdot d \geq 10$ cm

Height of beams $h \approx \frac{l_i}{17}$

IPE girders:

_ Load around the strong axis
_ Where h = section height in cm, q = distributed load in KN/m, l = span in m

$$h \approx \sqrt[3]{50 \cdot q \cdot l^2} - 2$$

HEB girders:

_ Load around the strong axis
_ Where h = section height in cm, q = distributed load in KN/m, l = span in m

$$h \approx \sqrt[3]{17.5 \cdot q \cdot l^2} - 2$$

Wide-span roof-bearing structures

Laminated timber beams (parallel):

_ Span 10–35 m
_ Distance between trusses 5–7.5 m

Height $h = \frac{l}{17}$

Timber trussed beams with parallel chords:

_ Span 7.5–60 m
_ Distance between trusses 4–10 m

Overall height $h \geq \frac{l}{12}$ to $\frac{l}{15}$

Steel solid web girder:

_ Span up to 20 m
_ IPE girder up to 600 mm high

Girder height $h \approx \frac{l}{30} \cdots \frac{l}{20}$

Steel trussed girder:

_ Span up to 75 m

Girder height $h \approx \frac{l}{15} \cdots \frac{l}{10}$

LITERATURE

James Ambrose: *Building Structures*, 2nd edition, John Wiley & Sons 1993

James Ambrose, Patrick Tripeny: *Simplified Engineering for Architects and Builders*, John Wiley & Sons 1993

Francis D.K. Ching: *Building Construction illustrated*, 3rd edition, John Wiley & Sons 2004

Andrea Deplazes (ed.): *Constructing Architecture*, Birkhäuser, Basel 2005

Heino Engel: *Structure Systems*, Hatje Cantz, Stuttgart 1997

Thomas Herzog, Michael Volz, Julius Natterer, Wolfgang Winter, Roland Schweizer: *Timber Construction Manual*, Birkhäuser, Basel 2003

Russell C. Hibbeler: *Structural Analysis*, 6th edition, Prentice Hall Publisher 2005

Friedbert Kind-Barkauskas, Bruno Kauhsen, Stefan Polonyi, Jörg Brandt: *Concrete Construction Manual*, Birkhäuser, Basel 2002

Angus J. Macdonald: *Structure and Architecture*, 2nd edition, Architectural Press 2001

Bjørn Normann Sandaker, *The Structural Basis of Architecture*, Whitney Library of Design, New York 1992

G.G. Schierle: *Structure in Architecture*, USC Custom Publishing, Los Angeles 2006

Helmut C. Schulitz, Werner Sobek, Karl-J. Habermann: *Steel Construction Manual*, Birkhäuser, Basel 2000

PICTURE CREDITS

Figure page 8: Colonnade in front of the Old National Gallery, Berlin, Friedrich August Stüler

Figure page 34: AEG Turbine Hall, Peter Behrens

Figure page 58: Berlin Central Station, von Gerkan, Marg und Partner

Figure 7, left, right; Figure 41, left, centre; Figure 55, left, right: Institut für Tragwerksplanung, Professor Berthold Burkhardt, Technische Universität Braunschweig

All other figures are supplied by the author.

P9

荷载和力

P9

承重结构和静力学

关于设计如何与施工联系起来的问题可以带来很多的思考。尽管可以从不同的角度去考虑，但通常它们也只是同一枚硬币的两个面。设计空间意味着通过将理论运用于实践从而实现它们。了解结构已成为建筑理论的基础之一。对于建筑师来讲（与结构工程师不同），很少能够保证施工建造中的稳定性，但他们应该能够在设计的初期阶段正确地选择结构构件并判断它们所需的实际尺寸。下一步通常是结构工程师建造承重系统。为了使构件能有效地组合在一起工作，建筑师应该掌握关于承重系统和结构、优缺点以及作用在上面的力的基本知识。这些不同的力初看起来很复杂，但本质上却是条理分明的。

解释它们是如何组合在一起的最简单的方法是按照进行静力计算的顺序理解。这种类型的计算通常按以下步骤进行：

—分析整个结构和其中单个构件的功能——建立静力系统；

—确定结构构件上作用的力——假定荷载；

—计算作用于特定结构构件的力以及传递给其他构件的力——计算外力；

—计算构件自身内部的力——确定内力或静力；

—确定设计构件的稳定性；

—验证设计构件可承受的力。

P9

力

$F = m \cdot a$

牛顿

力被定义为质量与加速度的乘积。

千牛：
1kN =1000N，
兆牛：
1MN =1000000N

力的单位是牛顿；1 牛顿对应的是大约 100 克的重量。在建筑中，其他力的单位还有千牛和兆牛。

一个力由其大小和方向所确定。它的作用是线性的（沿直线作用），可由作用线及作用线的方向所表示（图 1）。

力矩，扭矩

力也可以绕一点作用，叫做扭矩或力矩，由力的大小和力到杠杆支点的距离（力臂）的乘积确定。

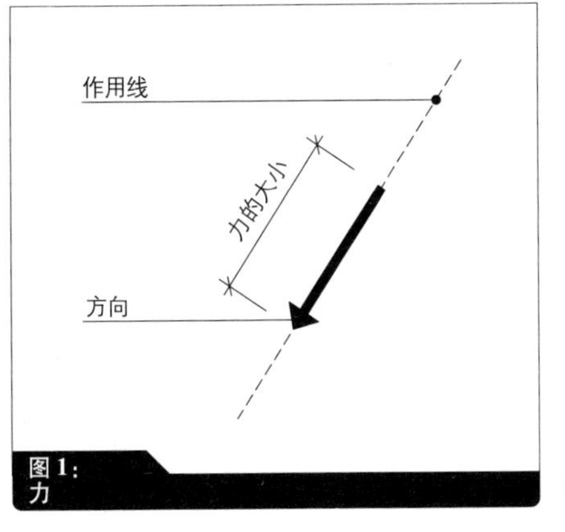

图 1:
力

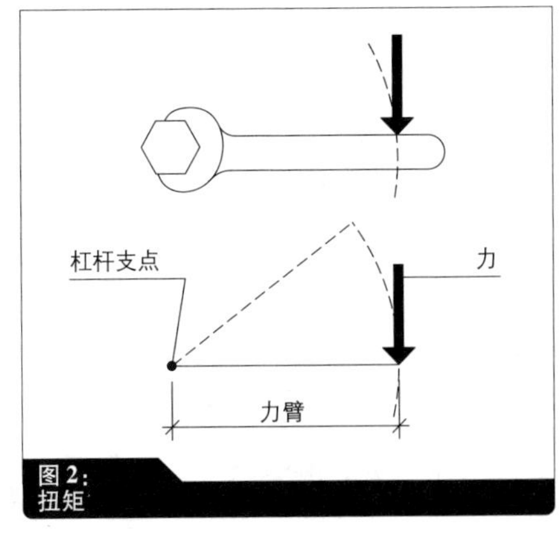

图 2:
扭矩

扭矩的简单例子就是用扳手拧紧螺丝。这也可以体现出力的幅值与力臂间的关系。力臂越长，扭矩越大（图 2）。

作用力 = 反作用力

静力学描述的是静止系统内力的分布。建筑或建筑的一些部分通常是不动的，其中所有的有效力都能够相互平衡。这符合“作用力 = 反作用力”定律。任何一个方向的力与相反方向力的总和为零是静力计算的基础。如果作用力已经知道，则反作用力能够马上确定。“外力，支撑力”一节介绍了将该理论应用于承重系统的方法。

P10

静力系统

结构工程师首先要在静力系统内建立结构间的联系。静力系统是构件所组成的实际复杂系统的抽象模型。尽管截面是有宽度的，支撑构件也仍然被认为是线条，而上面的荷载可以看作一个点。墙体是作为平面结构，荷载作用在线上。静力系统所能提供的其他信息是结构构件是如何连接在一起的，它们的力是如何在构件间传递的。这对于计算非常重要。静力系统中所用的符号将在“外力，支撑力”一节中说明（图 8，92 页）并应用于后面的章节。

位置

下面的工作是按顺序确定所有构件的位置并进行编号。在这里确定结构构件如何作用于其他构件也是非常重要的。

荷载传递

例如，屋面瓦不光是由屋面结构支撑，同样也会影响到墙，并传到基础。对于承受上层结构传递荷载的构件要非常精确地确定（图 3）。

›✎

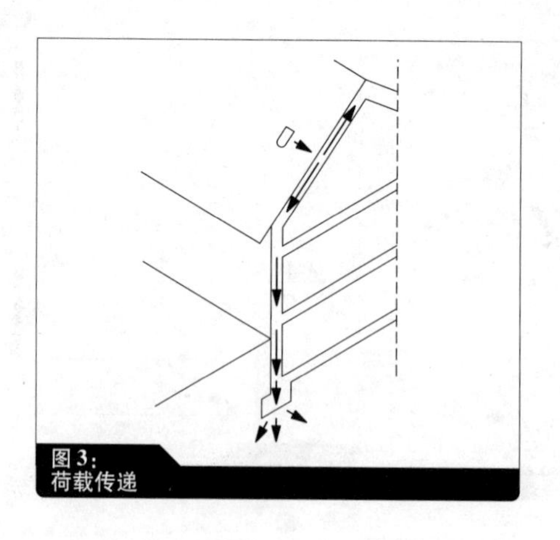

图3：
荷载传递

P11

外力

如果我们考虑一个建筑构件如屋面梁，我们要区分两种力。第一种是由上面的屋面结构施加，并经由屋面梁传到墙的力。如果我们不考虑它的自重恒荷载，则梁厚或薄、弱或强都是无关紧要的，因为我们现在处理的是外部作用力，它是不包括梁本身的。

我们必须要区分作用在梁上的内力和外力。例如，屋面梁为承受的屋面结构荷载而产生的弯曲力有多大？这个弯矩就是内力之一，将在以后对应的章节中说明。

作用力

所有对结构构件产生影响的因素都叫作用力。作用力通常是不同原因产生的力。从力学角度讲，影响构件的力又叫荷载。

注释：

为了与结构工程师很好地合作，设计师熟悉在一个项目中各个专业的组成部分，了解他们的工作方法和目标是非常重要的。仔细考察工程师的计算结果和工作计划并将其与建筑师的文献进行比较是有意义的。在结构工程师与建筑师设计完结构后，他们的主要工作是拟定平面布置允许的静力系统且为施工画出图纸。这里最重要的是建筑的承重部分。所有的非承重结构，例如非承重墙，都不会成为建筑布置的关键因素。

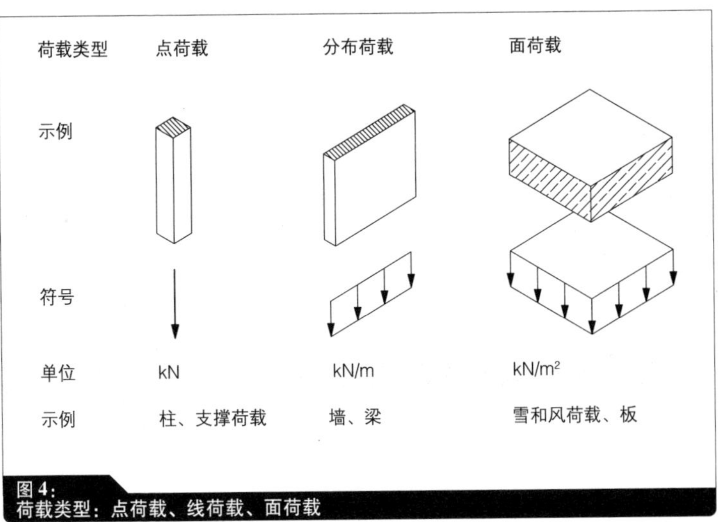

图 4：
荷载类型：点荷载、线荷载、面荷载

荷载

荷载从外部影响构件，我们必须将其与“支座反力”一节中提到的反作用力区分开。荷载可以分为很多种。根据静力系统简化程度的不同分为点荷载、线荷载和面荷载（图 4）。

另外，根据作用力持续性的不同，又分为恒作用力、可变作用力和异常作用力。

永久荷载

需要注意的是恒作用力包括构件重力，称为永久荷载（恒荷载）。

工作荷载

工作荷载包括风、雪和冰等不同的作用。工作荷载需按标准级别设计，从而满足建筑使用要求。最重要的是作用在楼板上的竖向工作荷载。不管房屋是作为住宅、办公室、会议室或其他用途，都必须要选择合适的工作荷载值。较大的水平荷载也需要考虑，如栏杆、女儿墙、汽车的刹车、加速和碰撞、机械的动力作用和地震作用等。这些荷载的大小在德国国家规范中已经列出（见“附录”，“参考文献”）。

假定荷载

在利用静力系统解释了结构如何工作后，下一步就是确定作用力。所有的作用力必须被识别出来、确定取值并组合在一起。它们通常是与结构构件的长度和面积相关。斜向作用的荷载通常被分解为水平方向和竖直方向的分力。

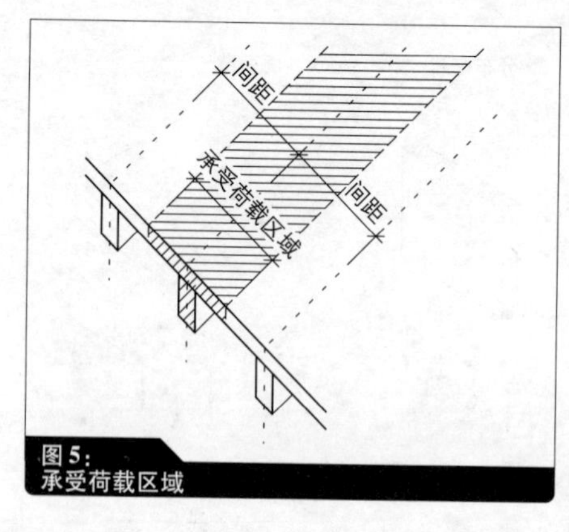

图 5：
承受荷载区域

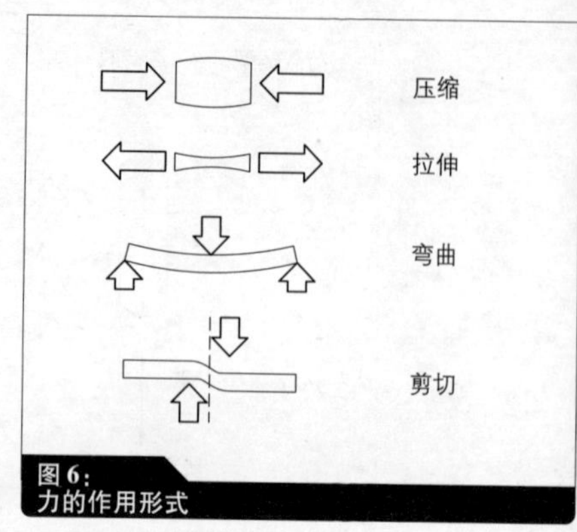

图 6：
力的作用形式

竖向荷载
水平荷载

承受荷载
区域

为了进一步的计算，我们需要区分竖向荷载、水平荷载和扭矩。

承受荷载区域描述了荷载作用在结构构件上时对其产生影响的特定区域，是承受荷载的某个结构构件表面的一部分，与结构的材料和跨度相关。

例如木楼板梁间隔 80cm，对于单独的一根梁来讲，哪一部分楼板荷载作用其上？荷载影响区域是从左手边梁间区域的中心线直到右手边梁间区域的中心线，即 40cm 的两倍，所以它的总宽度为 80cm（图 5）。这是个简单的例子。对于特殊的结构构件而言，确定其荷载影响区域将十分复杂。

力的作用形式

到现在为止，我们已经认识了荷载及其幅值。但了解一个荷载，或更一般地说，一个力如何作用在结构构件上也是十分重要的。在这里我们将区分不同的作用形式。

—受压：一块石头放在另一块石头上，对下面这块石头施加的就是压力。

—受拉：如果用绳子来举例说明拉伸荷载是最清楚的，绳子只能承受拉力。

—受弯：梁两端固定，从上面加载，梁将下弯，也就是说，它承受了弯曲荷载。

图7：
钢结构中的支座

—受剪：这可以用家里剪刀剪纸来解释。两个力错开一点距离沿相反方向横向作用在结构构件上。这种荷载通常作用在连接装置如螺栓上（图6）。

支撑

传递荷载的两个结构构件的连接部分叫做支撑。举一个简单例子，屋面梁就是依靠支撑架立在石墙上。在建筑中，支撑的概念更加广泛，包括许多不同的构件连接部分。例如，当一根旗杆固定在地面上或者一根钢梁与柱连接时，这就叫做一个支撑。从结构工程来讲，支撑主要是因为所承受力的不同而有所区别。

在旧的钢桥中，可以很容易看到不同的支撑形式。巨大的梁支撑在非常小的点或狭窄的区域上，这意味着梁的弯曲不必受到支座的干扰，这也就是所说的“铰接支座”。铰接支座在梁的一端，另一端则支撑在钢辊上。

重要提示：

在结构构件内单位面积竖向作用：自重，楼板、楼梯和阳台的工作荷载。

在楼板上单位面积竖向作用：雪荷载。

在结构构件内单位面积以一定角度作用：风荷载。

水平作用：作用在女儿墙和栏杆上的荷载，刹车和加速荷载，汽车的碰撞荷载，地震荷载。

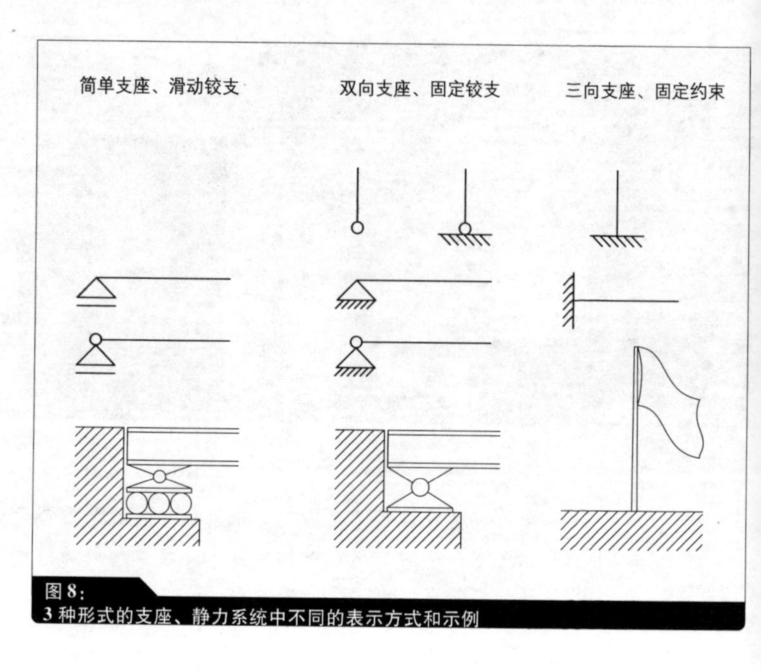

图 8：
3 种形式的支座、静力系统中不同的表示方式和示例

滑动支座

当梁受热伸长，支座将在这些钢辊上移动，从而弥补长度上的差异。这种支座可以抵抗竖向力，但是无法抵抗由温度变化所引起的水平力，并且它们同样也不限制梁的弯曲，被称为滑动支座。

固定支座
铰支支座

没有支撑在钢辊上的支撑可以同时传递竖向力和水平力。它们被称为固定支座或铰支。

约束

下面再考虑上面提到过的固定在地面上的旗杆。它的锚固可以传递从旗杆到地面的竖向力和水平力，同时防止旗杆倾覆（绕支撑的转动）。这种支撑叫做约束（图 8）。

我们可以区分以下三种支撑：

—简单支撑。只能抵抗一个方向的力，可以滑动的铰接。

—双向支撑。可以抵抗几个方向的力，固定的铰接。

—约束。是三向支撑，可以抵抗不同方向的力以及弯矩。

在建筑中，选择恰当的支撑是非常重要的，并且必须在静力系统中表示出来。

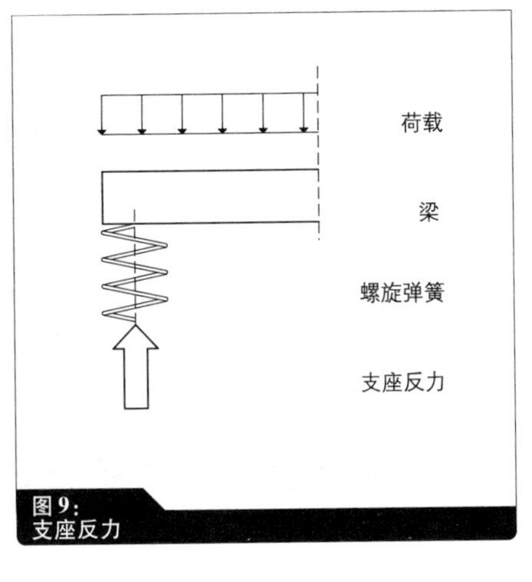

图 9：
支座反力

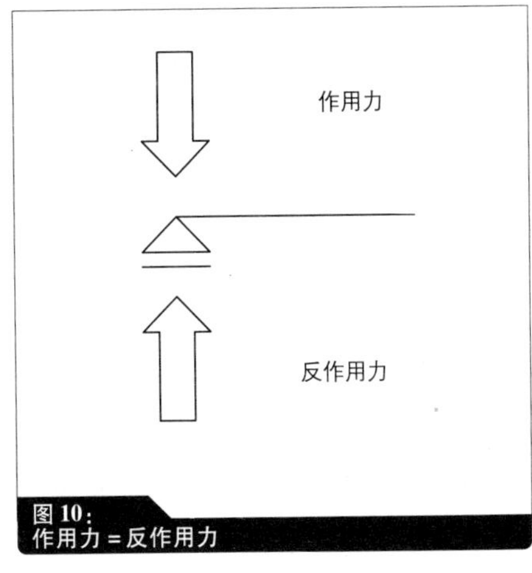

图 10：
作用力 = 反作用力

支撑力

让我们假设梁不是由石墙而是由弹簧支撑，弹簧被梁上传下的荷载挤压并产生一个反力。

支撑反作用力

这个反力称之为支撑反作用力（图 9）。如果梁不移动，则弹簧反力与梁施加的作用力大小相同。简单地说，就是作用力等于反作用力（图 10）。这在起支撑作用的石墙中是不可能用肉眼看到的，但是石墙如同弹簧一样受到挤压，因此它也产生支撑反作用力。

当计算建筑时，一定要知道支撑结构构件的支撑力。在确定作用荷载后，支撑力通常能够很快地计算出来。应用上面提到的"力 = 反作用力"法则，能够建立三个这样的等式来计算支撑反力。这三个法则是进行静力计算的基础。

平衡方程条件

它们被称作平衡方程的三个条件（图 11）：

$$\Sigma V = 0$$

所有的竖向荷载总和与竖向支撑力总和相等。这意味着：竖向力总和为 0。

$$\Sigma H = 0$$

所有的水平荷载总和与水平支撑力总和相等。这意味着：水平力总和为 0。

$$\Sigma M_{\mathrm{p}} = 0$$

考虑在支撑点 P 处的支撑，所有绕该点沿顺时针方向转动的力与沿逆时针方向转动的力相等。这意味着：绕给定点的所有弯矩总和为

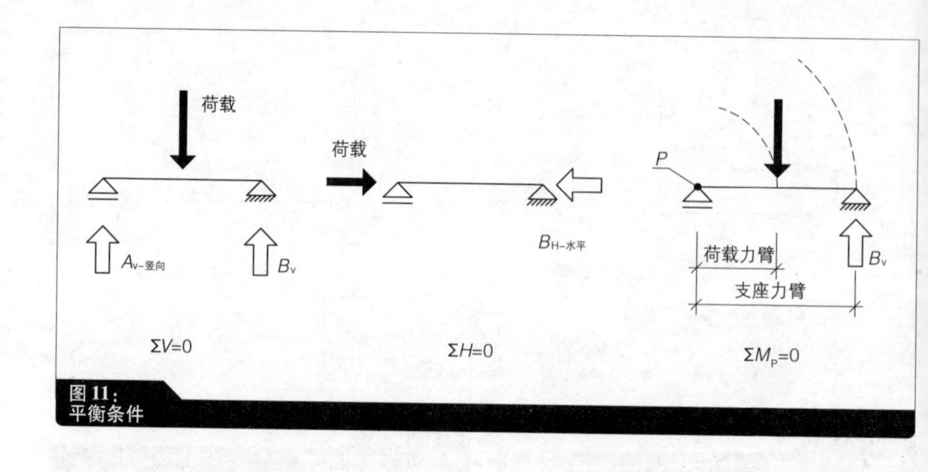

图 11：
平衡条件

0。这里需要指出的是所有力和荷载都可以看作是绕固定点转动的扭矩。根据定义，力乘以力臂等于扭矩的大小（参见“力”一节）。

只有得到绕一个支座的所有弯矩，才能计算出两个支座的支撑反力。因为扭矩中心是支座，因此，该支座的反力没有力臂，在平衡方程中力矩为0。这意味着在平衡方程中只有一个未知量，即另一支座的反力，这可以很容易地计算出。

图 11 所示有中心荷载作用的梁，绕 P 点旋转的扭矩方程为

$$\Sigma\widehat{M}_A = 0 = A_V \cdot 0 + F \cdot l/2 - B_V \cdot l \rightarrow B_V = \frac{F \cdot l}{l \cdot 2} \rightarrow B_V = F/2$$

两个支座分别承担一半的中心荷载。在这种情况下，不必计算就可以得到这样的结论。

在利用平衡方程条件进行计算时应该确定符号规则。如果符号规则未定义，就必须采用箭头表示力。它显示力的正方向。在这里向右的力为正，因此，向左的力需要用负号注明。

P18

内力

到目前为止，我们只讨论了作用于结构上的力以及相应的支座反力。这些被称之为外力，因为并未考虑结构构件自身。但在梁内部发生了什么？或者换个方式说，构件内哪个力是有效的？

为理解这些，假设支撑在两个支座上的梁在任意位置被截断。将会发生什么？梁将倒塌，无法支撑任何东西，甚至包括自己。现在的

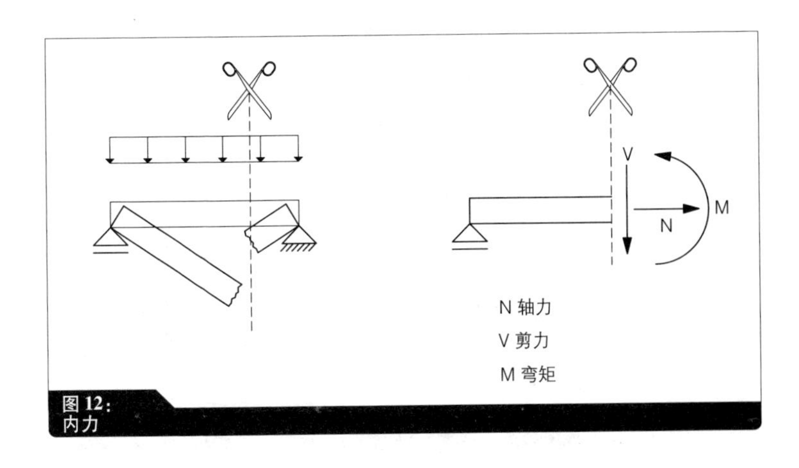

图 12：
内力

问题是在断面上，哪个力将起作用，防止梁倒塌，或者说，为了实现内力的平衡需要哪个力？

在这里上面提到的平衡条件是有效的，它们可以同样用于内力和外力平衡。假设作用于截断面到支座的外力与截断面上产生的内力大小相等（图 12）。

内力

正如同外力分为竖向力、水平力和扭矩一样，内力也分为轴力、剪力和弯矩，其方向与结构构件自身相关。

轴力

轴力是沿纵向或构件方向作用的力。我们可以用一根绳子挂在钩子上，绳下面悬挂一个重物的例子来解释轴力（图 13）。重物就是荷载，钩子提供支撑反力。这些是外力。

轴力

如果不考虑绳子自身的重量，在绳子中，各点拉力是相等的。在这里，绳子的长短没有任何影响。所以在绳子内各点存在相等的轴力，大小等于悬挂的重量。

在“力的作用形式”一节中，解释了两种纵向作用的力：压力和拉力。这就是轴力。

另一个例子是无侧向支撑的石墩（图 14）。仅考虑墩的自身恒荷载（石头是一种自重很大的材料）。很容易计算出桥墩底部基础的支撑反力是与桥墩的重量相等的。但是在桥墩内部呢？最上面的石头并未受荷载作用，所以在这点上没有轴力。上面数第二块石头承担最上面

压力

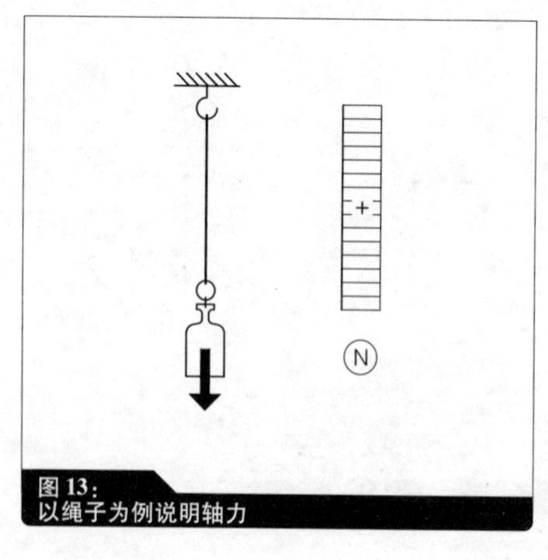

图 13：
以绳子为例说明轴力

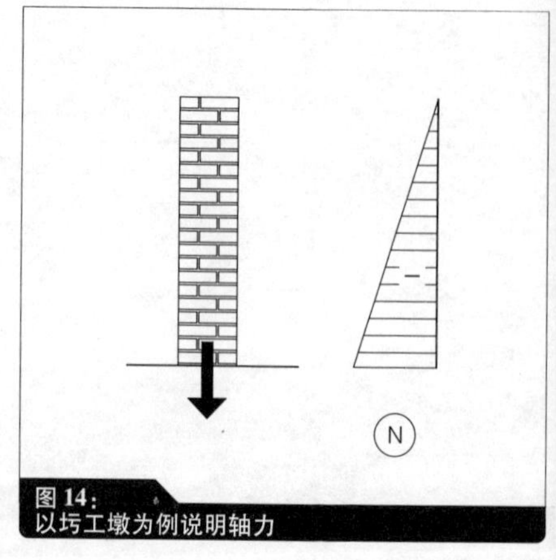

图 14：
以圬工墩为例说明轴力

石头的重量。因此在上面数第二块石头这点有一个小的压力。越向底部，压力逐渐变大。也就是说，轴力在桥墩内自上而下不断增加（见“力”）。

如同荷载一样，轴力的大小可以用图表示。以上的两个例子说明了不同的轴力。在图中，拉力用正号表示，压力用负号表示（图 13、图 14）。

剪力

对于外力来说，水平力和竖向力是有区别的。内力之间也有区别，它们的方向与各种情况下构件的系统轴相关。如同沿纵向作用的拉力与压力被定义为轴向力一样，所有沿横向作用的力被定义为剪力。它们不像轴向力一样容易理解，一定不要与下面一节的弯曲混淆。

悬臂梁

以一个悬臂梁为例解释剪力效应，图 15 为一端固定在墙上的梁。这种梁叫做悬臂梁，例如，阳台的一部分受到自重影响（可以认为受到均布荷载作用）。如果梁在靠近端部被截断，由于均布荷载作用，截下的这段会掉落。荷载沿轴的横向作用并产生剪力。如果更长的一段被截断，那么在截断点上需要承担更多的沿构件横向作用的均布荷载。因此这种情况下该点的剪力大于上面的那种情况。随着截断长度的增加，力也随之增加。于是剪力从自由端向固定端增加。因此在固定端支撑力与剪力相等。

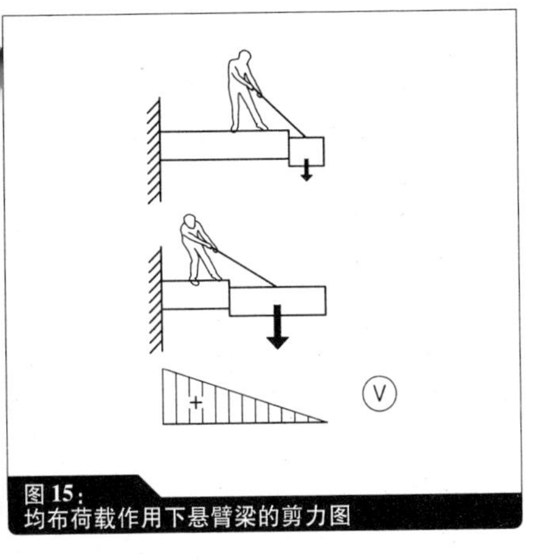

图 15：
均布荷载作用下悬臂梁的剪力图

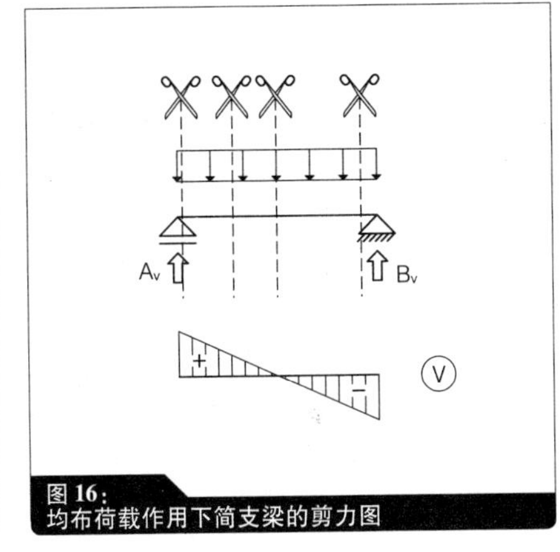

图 16：
均布荷载作用下简支梁的剪力图

简支梁

图 16 是一根支撑在两个支座上的梁，叫做简支梁，作用着均布荷载。设想从左至右依次截断截面，并且考虑在截面的左边作用着什么外力，就会很容易理解剪力的作用。

第一个有意思的截断面是左支撑的右截面。其上发生了什么？支撑反力沿横向作用，方向向上，因此在该截面上产生了剪力。但如果截断面向右发展，一部分线荷载将与支撑反力反向作用，因此与前面情况相比，剪力减小。

现在在梁的中部截开。在梁左端至截断面处沿横向作用着哪些力？首先是向上的支撑反力，其次是左端到中部的均布荷载。因此整根梁上一半的均布荷载是有效的。在这样的对称系统中，很容易判断每个支撑承担一半的均布荷载。因此在这里梁中部的剪力为 0。

如果我们现在考虑一个右边的截断面。更大的一部分线荷载发挥作用，这意味着剪力已经为负。在左支座前面的截面上，几乎全部的分布荷载与左支座几乎未变的支撑反力相作用。只有算上右支座的支撑反力，总和才能相加为 0。

如果考虑梁的右截面而不是左截面，结果是相同的。因此，无论考虑哪个子系统都没有关系，因为梁内每点的内力都处于平衡状态。这对所有的内力都适用。

弯矩

弯矩的效果在“外力”一节已经讨论过。在这里所有的有效力都看作绕着固定的一点旋转。它们的幅值定义为力与力臂的乘积。（“力”、“外力”、“支撑力”等节）与外力会影响支座反力一样，作用在梁上的力对于确定其内部弯矩来说是非常重要的。

弯曲

内部弯矩使梁发生弯曲。弯曲是确定构件尺寸的关键荷载。在进行静力计算时，知道梁任意点上的弯矩的大小是非常必要的。这可以在弯矩图中表示出来。弯矩图是设计建造在弯矩作用下构件的重要辅助手段。

内部弯矩同弯曲荷载间的联系将依旧用悬臂梁在下面进行说明。在均布荷载作用下，悬臂梁如何变形？荷载导致梁向下弯曲（图 17）。这里，弯曲变形意味着梁的上部变长而底部变短，从而在上方的伸展面上产生拉力，而在底部的压缩面上产生压力。这种张力是由荷载产生的内力。

弯曲自身产生内部弯矩，其幅值与外力幅值及力臂长度相关。对于悬臂梁，在自由端均布荷载大小和力臂长度都很小，因此弯矩很小。但在固定端，全部的均布荷载都是有效的，且力臂长度很大，因此弯矩也很大（图 17）。

在单个集中荷载作用下，简支梁将向下变形。因此，与上面的悬臂梁相反，简支梁的底部受拉而顶部受压。

与上面的悬臂梁例子相比，弯曲沿不同的方向作用。这如何影响力的分布？让我们从左到右对其进行分析。

支撑力作用于左面支座的右截面上，但是力臂长度为 0，因此力矩为 0。力臂长度随到支座的距离增加而加大，弯矩呈线性增加，一直到外荷载的作用点。再向右，外荷载作用方向与支座反力相反，而力臂长度增加，弯矩减小直到在右支座处为 0。我们可以分别从左或从右进行检验，其结果是相同的（图 18）。

当均布力 q 而不是一个集中荷载作用时，力的大小及分布如何变化？均布荷载可以合成为一个作用在均布荷载重心处的集中荷载。集中荷载的幅值为力的单位长度乘以其有效长度。

$R = \frac{q \cdot l [kN \cdot m]}{m}$

为计算弯矩，必须在不同截面上依次确定合成集中荷载大小及其力臂长度（图 19）。这些集中力与支座反力方向相反，且幅值增大。弯矩图为一条抛物线，因为在计算分布荷载大小及力臂长度时长度被考虑了两次。

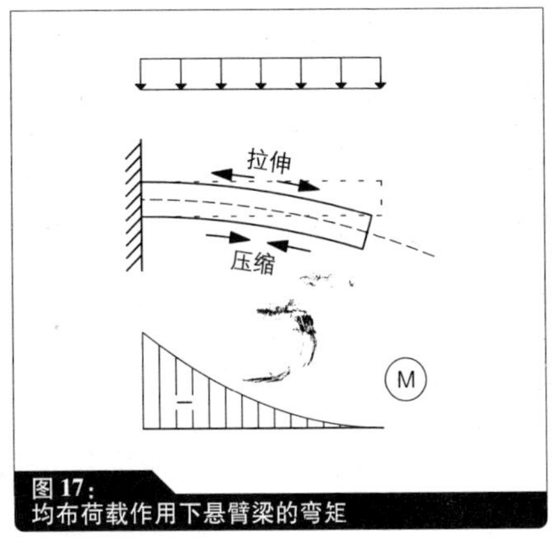

图 17：
均布荷载作用下悬臂梁的弯矩

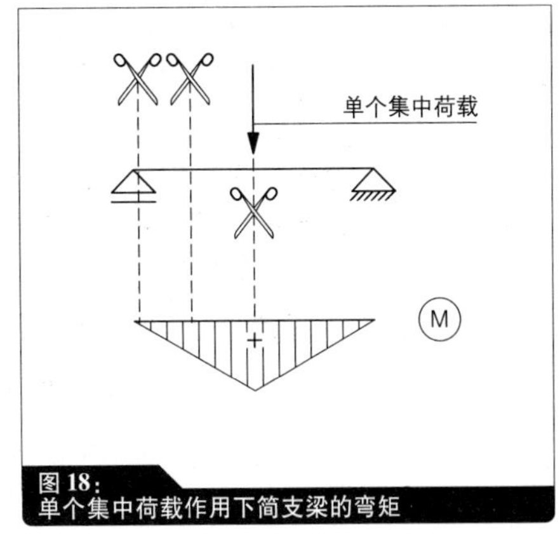

图 18：
单个集中荷载作用下简支梁的弯矩

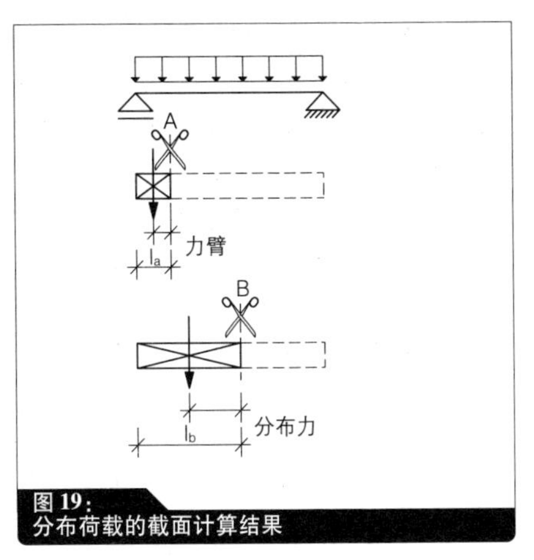

图 19：
分布荷载的截面计算结果

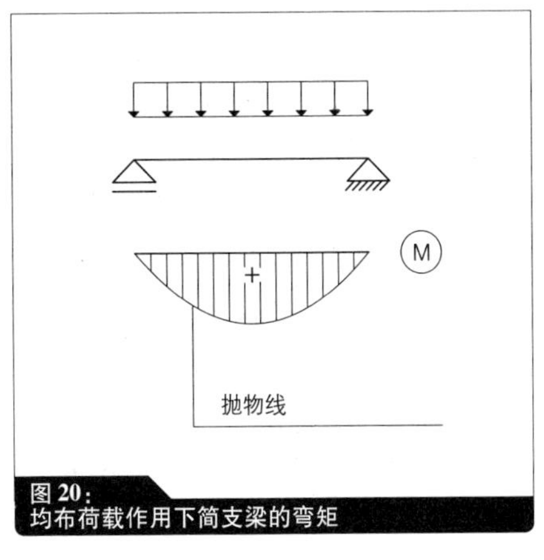

图 20：
均布荷载作用下简支梁的弯矩

分布荷载弯矩：$M_A = q \cdot l \cdot l/2 \rightarrow M_A = \dfrac{q \cdot l^2}{2}$

对于弯矩图来讲，支座位置是非常重要的点，弯曲力在支座处为0。这如何解释？如果我们在支座处截开且向支座方向看（图 19），没有任何一个力具有力臂，因为构件在这里没有可测量的长度，实际上我们是在考察一点。一般来讲，弯曲需要一个固定的梁截面抵抗弯矩。但在铰节点处是不适用的。例如，一根链子是铰节点的组合，因此无法承受弯曲作用。因此我们有一个很重要的原则：铰节点处弯矩为零（图 20）。

最大弯矩

在本例中，最大弯矩出现在跨中处。为承担荷载，梁必须能够抵抗这个最大弯矩。一般来说，当决定一根构件的尺寸时，应首先确定弯曲力、作用位置及最大弯矩值。

当设计一根复杂的大跨度梁时，最大弯矩并不是惟一的重要参数。根据材料性质设计符合弯矩图的梁截面才是经济的。换句话就是说，设计出的梁应使其尺寸与每点的有效弯矩精确匹配。为此，建筑师应根据梁所受荷载定量地确定弯矩图。

内力之间的联系

上面介绍了三种不同的内力。当计算承载结构时，确定所有三种内力是非常必要的，在此基础上，能够确定三种内力共同作用下的结构尺寸。

剪力与弯矩联系紧密，二者由相同的荷载引起，可以互相推导出来。例如，如果没有力绕着杆作用，剪力值不会发生变化，即为常数。但是因为弯矩定义为力与力臂的乘积，则其幅值变化在无荷载区域是线性的。如果一个力作用在一个特定位置，则弯矩值与该力作用位置的距离成正比。这种剪力图和弯矩图间的关系是固定的（图 21）。

下面一些特有的关系对于静力计算是非常重要的：如果我们比较力的图形，我们可以发现在力矩最大处，剪力总是为 0。这条结论非常有用，因为弯矩最大值的位置可由剪力图确定，且只能在该点进行计算（图 21，图 22）。

可以通过经验定量地确定剪力图和弯矩图。图 23 显示了一些常用荷载类型的力的图形。

提示：

表示内力的符号如下所示：

轴力：压力（－）代表向上，压力（＋）代表向下。

剪力：正的剪力画在系统线上，负的在下。

弯矩：弯矩画在弯曲方向，正弯矩向下，负弯矩向上。

但是这些惯例不要认为是固定的。例如有些国家是以别的方式定义弯矩的。

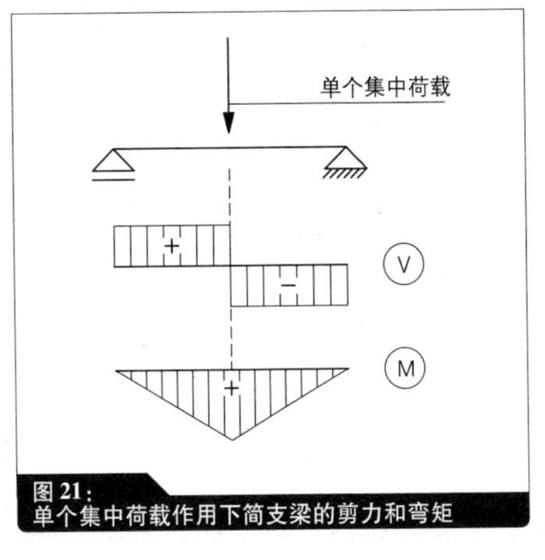

图 21：
单个集中荷载作用下简支梁的剪力和弯矩

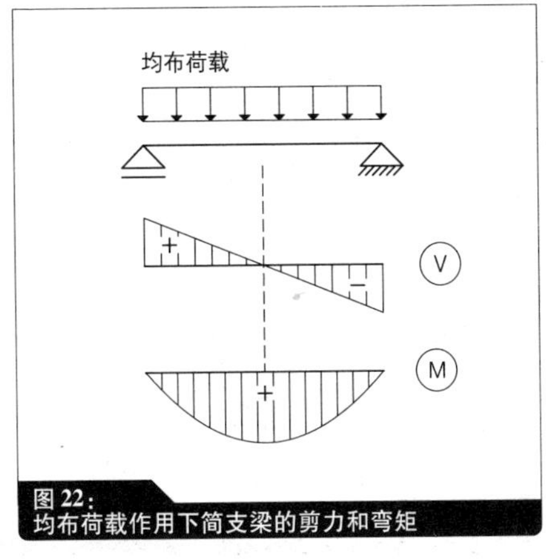

图 22：
均布荷载作用下简支梁的剪力和弯矩

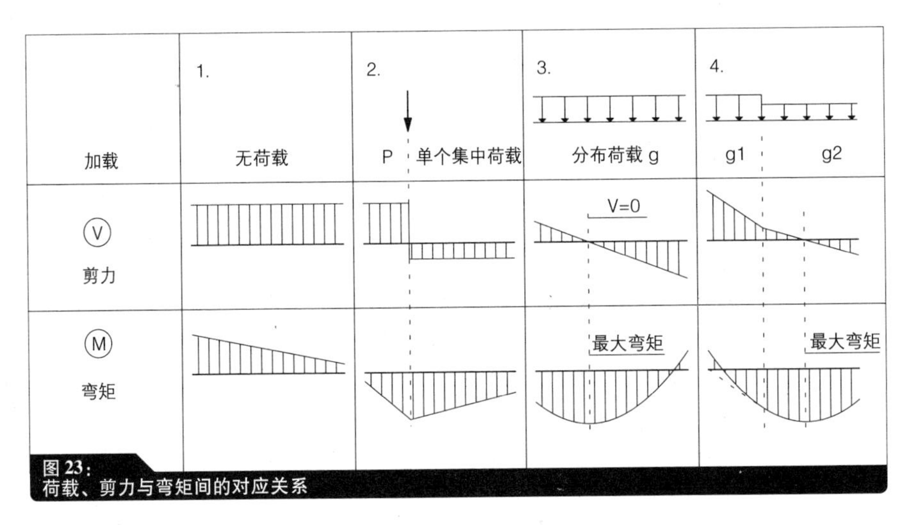

图 23：
荷载、剪力与弯矩间的对应关系

1. 如果没有力作用在杆件上，则剪力为常数，而弯矩线性变化。

2. 集中荷载使剪力产生间断，并使弯矩图发生弯角。

3. 对于均布荷载，剪力图为一斜直线，弯矩是抛物线。

4. 均布荷载的间断在剪力图上产生弯角；在弯矩图上，两条不同斜率的抛物线组合在一起，并且具有相同的切线。表中栏 1 和栏 2 是图 21 中系统的详细表述，图 22 是栏 3 的例子。

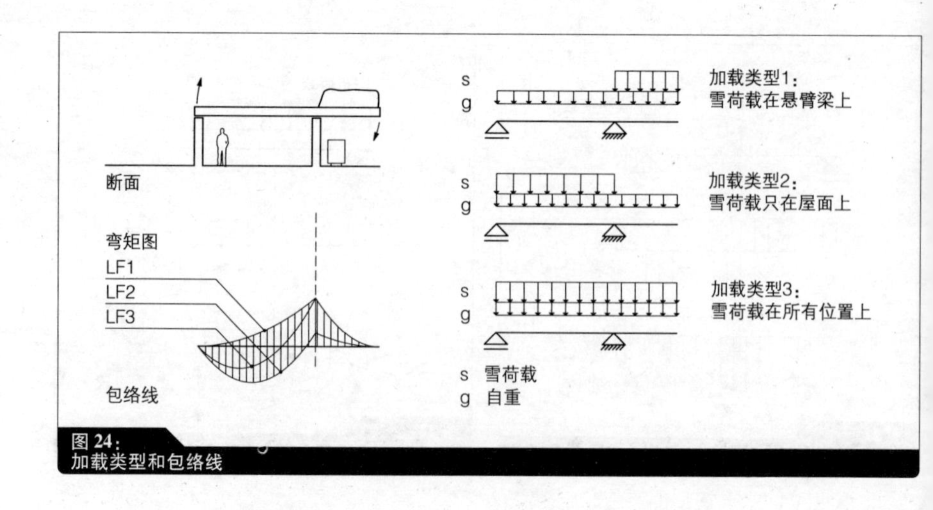

图 24：
加载类型和包络线

加载类型

实际上，许多不同的荷载是叠加作用的。在计算中，它们应该加在一起从而决定承担最大荷载的构件尺寸。但是也有这种情况，并不一定是最大荷载才会产生危险。内力的最大值是决定构件尺寸的关键因素，同样也可能是其他荷载的组合成为关键因素。这些荷载的不同组合叫做加载类型。

考虑下面的例子：一间小厂房有平的木屋顶，在一侧有突出的顶棚以便储存材料。冬天有很大的降雪。厂房的热供应很好且屋面的隔热较差，则屋面的雪开始融化，但顶棚的雪却没有融化，因为下面是没有供暖的区域。这样的荷载导致屋面的危险增加，与突出部分相邻的屋面梁可能发生破坏（图 24）。

为了避免这些危险，结构工程师不仅要将建筑的雪荷载作为整体计算，而且要单独计算作用在顶棚上的雪荷载，因为它们产生不同的危险。第一步是确定可能有什么样的加载类型或组合，然后将它们组合在一起。如果可以分别画出不同荷载类型的弯矩图，则可以从图中很容易地判断出各点可能的荷载最大值。

包络曲线

该图也叫做包络曲线。它的极值点显示了对各点非常关键的加载类型。图 24 显示了梁跨内最大的正弯矩在荷载工况 2 内出现。但最大的负弯矩在工况 1 和 3 内出现。

P28

确定尺寸

静力计算过程与上面几章提到的计算顺序是一致的。先建立静力模型后，计算假定荷载，然后是计算外力，最后是确定结构构件的内力。

如果我们现在能够简单地计算出构件所需截面当然是很好的。但不幸的是这不像看起来那样容易。因为如同荷载一样，结构的所有构件都已成为计算的一部分，换句话说，就是如果想计算出假定荷载，必须要知道所有的结构因素，甚至构件重量也必须要考虑在内。如果计算结果表明，结构的其中一根构件不能承受足够的荷载，则我们不得不一切从头再来。

即使如此，也并不证明所有的工作都是无用功，非常明显，事先仔细的规划是非常有利的：尺寸可以事先确定下来。这可以借助一些简单的原则完成（见“附录”，“确定构件尺寸的原则”）。

强度

作用力确定后，需要引起注意的就是结构构件的承载能力了。这主要取决于两个方面：材料和截面。

设计的第一步是确定材料。每种建筑材料都有自己的优点和缺点。对于建筑来说，不同材料所能提供的强度或抵抗力是非常重要的。例如，索结构可以承受拉力却无法受压，而砌体只能受压却无法受拉。木、钢和钢筋混凝土结构可以抵抗拉压作用，同时也可以抵抗内力（见“外力”，“力的作用形式”等节）。

前面已经说过弯曲可以同时产生压缩和拉伸作用。因此，当弯曲荷载作用时，只有同时承受拉压作用的材料才能被使用（例如木构件和钢构件）。

应力

$$\sigma = \frac{F\ [kN]}{A\ [m^2]}$$

罗伯特·虎克

(1635－1703)

材料在承载能力方面也有不同。该能力表现在指定面积上承受的荷载大小。每单位面积的强度用应力 σ 来表示。

为了理解这个受力的概念，我们引用虎克定律，即拉伸在弹性方面成正比。对建筑材料来说，这意味着什么呢？每种材料，无论木材、钢材、混凝土或是砌体，从根本上说，都是弹性的。如果一根建筑构件受力，产生的拉力与材料的拉伸变形成正比。所以当梁受载时，会产生弯曲或下垂，如果荷载加倍，则变形值是前者的二倍，如果荷载减小，则变形减少。

这种变形模式有一个上限，如果拉力过大，则材料不再发生弹性变形，而是进入塑性变形阶段，即产生不可恢复的永久变形。从这点开始，结构构件开始发生损坏。如果进一步加载，则会发生完全的破坏，尽管各种材料的破坏形式有所不同。材料在进入塑性阶段和破坏前能承担多少的荷载值完全是材料本身的一个性质，与结构构件的几何尺寸无关。重要的是，在建筑最大可能荷载作用下，材料没有达到最大允许值。实验室条件下，考虑不同的材料质量，可以确定特定材料所能承受的拉力。

允许拉力

这种方法确定的值叫做允许拉力，能够从表中直接得到（见“附录”，“参考文献”）。

强度等级

另外，不同质量的材料其允许拉力也不同，可以分为不同强度等级。例如，普通混凝土和高强混凝土可按强度等级区分。结构构件实际承载能力的确定通常基于以下原则：实际拉力应小于允许拉力。如果结构构件轴向加载，计算出这些拉力是很容易的。拉力与结构构件截面单位面积对应的轴力相等。如果结果显示拉力小于允许拉力，则该结构尺寸设计没有问题。但不幸的是，在确定结构尺寸时这种简单的情况非常少见。索因为只能承受拉力，所以可以用这种方式确定，但在其他大多数情况下，弯曲荷载才是确定结构尺寸的重要因素。

抗弯能力

结构构件的合适与否都取决于实际拉力值是否小于允许拉力。这对于受弯构件也同样适用。

压力分布

当解释弯矩时，我们曾经说过，受弯构件一面承受拉应力、一面承受压应力，但是这些应力有多大，是如何分布的呢？

为解决这个问题，我们考虑一个未承载构件，横向用直线作出标记。当构件受荷弯曲时，标记将相对倾斜形成梯形，但直线依旧保持为直线（图25）。如果拉伸和压缩是成比例的，就会形成如下的应力分布，从底部的拉应力，经过中线（没有拉应力），直到顶部的压应力，都保持直线。

中和压力面

由图26可见，拉力分布和压力分布都形成了一个三角形。这些应力三角形都可以变换为两个作用在三角形重心处的集中力，二者之间的距离为截面高度的2/3。这个长度就是内部弯矩抵抗外加荷载的力臂长度。所以构件高度越大，内部拉力的力臂越长，结构的稳定性越好。

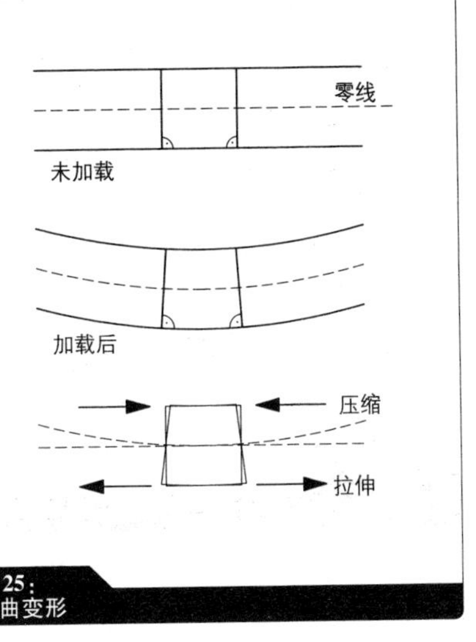

图 25：
弯曲变形

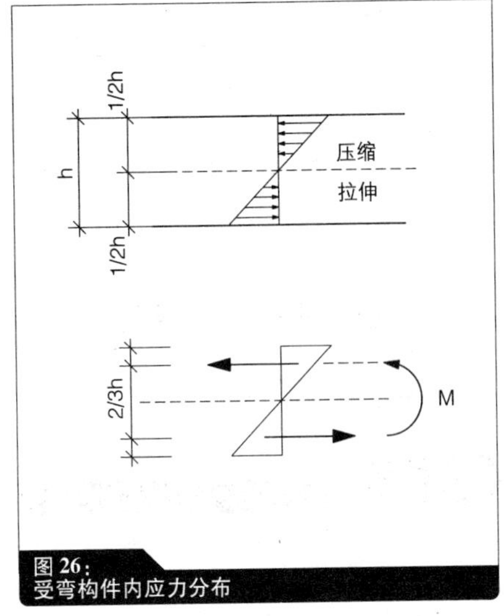

图 26：
受弯构件内应力分布

所以，力臂长度是抗弯的关键，同时构件宽度也是非常重要的。截面的抵抗力表示为抗弯能力。抗弯能力与构件的几何尺寸相关而与材料无关。

例如，木结构中常见的矩形截面的截面抵抗矩为 $W = w \cdot h^2/6$。

再深入分析一下这个公式：高度 h 是平方项，而宽度 w 只是一次项。一个立起来的矩形比正方形或是放倒的矩形有更高的承载能力。精确地说，将宽度加倍可以使承载能力加倍，而高度加倍将使承载能力增大 4 倍。

对于简单的矩形截面的抵抗矩可以利用上面的公式方便地得到，对于其他形式的截面，如钢构件的截面是非常复杂的。基于这个原因，截面抵抗矩的值总是以表格的形式给出（见“附录”，“参考文献”）。

“弯矩抵抗矩“这个名词包含弯矩这个词。与上一章“力”中弯矩或扭矩的概念不同，抵抗矩不是指有特定力臂的单个力，而是指绕拉伸为零的那道线的构件面积（图 26）。如同下一章中的惯性矩一样，弯矩抵抗矩也可以被定义为面积矩。

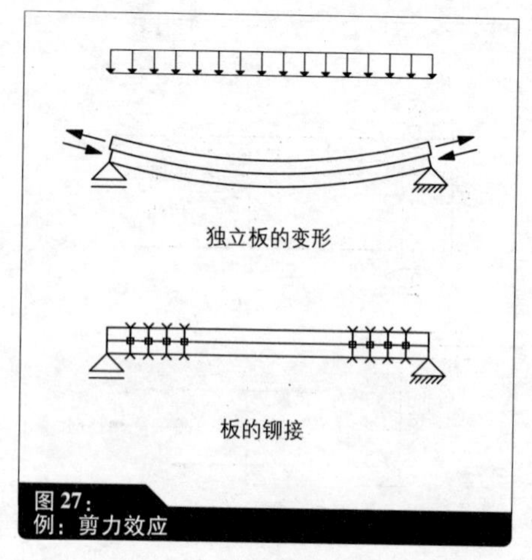

图 27：
例：剪力效应

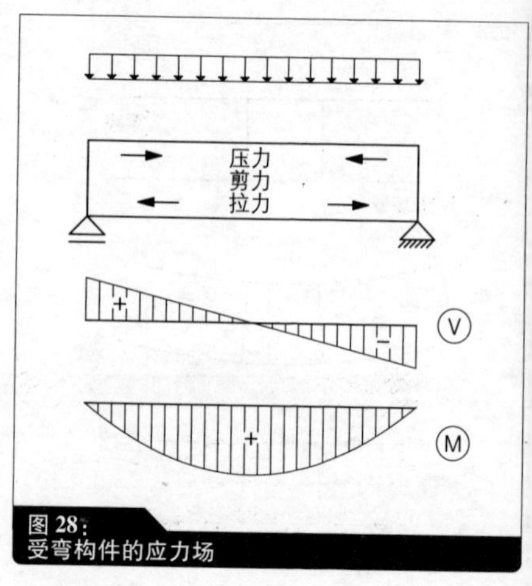

图 28：
受弯构件的应力场

惯性矩

惯性矩可以用它的效应很好地来解释。抵抗矩表示的是一个构件对弯矩的抵抗能力，而惯性矩与变形相关。它描述了截面的刚度。

像抵抗矩一样，惯性矩也是基于受弯截面上应力的分布。这里边缘的受压区域和受拉区域要比零拉力线区域更加有效。但是面积单元到零线的距离对于惯性矩比对抵抗矩重要得多。

惯性矩是截面上所有面单元的面积与它们到零线距离的平方的总和。

对于矩形截面来说，可以用公式 $I = w \cdot h^3/12$ 得到其惯性矩。所以这里截面高度已经到了 3 次方，意味着当构件高度加倍，宽度保持不变，则挠度减小为原来的 1/8。

变形

惯性矩可以用来计算构件的变形。尽管截面弯矩抵抗矩是确定承载构件几何尺寸的前提，但不超过最大允许变形值也是必要的。

剪应力

让我们以下面的例子进行说明。两块板叠在一起作为简支梁，在上面加载。两块板在荷载作用下都将发生变形，并相互错动（图 27，28）。它们可以牢固地连接在一起以提高承载力，因为高的截面比两块叠起来的板有着更大的承载能力（见“确定尺寸”，“弯矩抵抗矩和惯性矩”）。最好应该怎样处理？

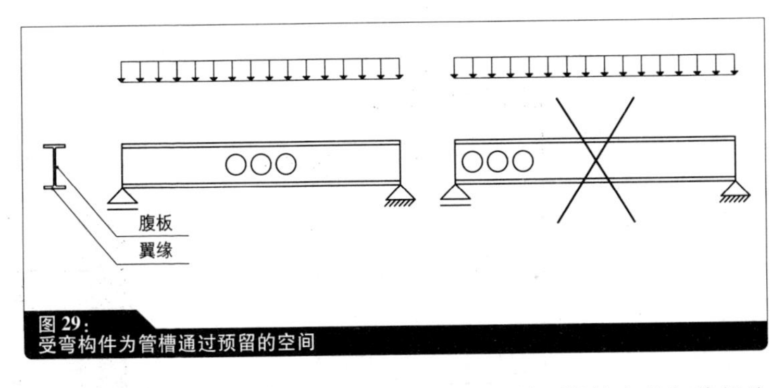

图 29：
受弯构件为管槽通过预留的空间

一种可能的方法是在未承载的板上钻孔，并用螺栓和铆钉将板连接起来。

现在我们需要了解这些铆钉承受着怎样的应力并发挥什么样的作用？问题第一部分的答案很简单。板的相互错动会产生剪应力。解释剪应力从何而来的最简单的方式如图 28 构件所示。

构件承受着均布荷载，最大拉应力在跨中下边缘，最大压应力在跨中上边缘。随着向支座方向弯矩的减小，应力也减小。但这些应力不可能会简单消失，那到底发生了什么？向支座方向拉压应力在逐渐减小，而随着弯曲应力减小，剪应力在增加。

在弯矩图中，可以识别出弯曲应力。而剪应力是与剪力成正比的。在构件承受均布荷载时，向支座方向，剪力逐渐增加（见“内力”，剪力）。

这种由弯曲产生的拉压应力最大处位于梁跨中的上下边缘（见“内力”、弯矩）。最大剪应力在支座处。例如木材是一种对剪应力很敏感的材料。在木结构中，对支座处抵抗剪力的木材进行加强是很必要的。

工字梁截面

我们可以举另一个例子进一步认识剪应力。常用的钢构件截面，如工字梁的翼缘可以承受弯曲，而腹板承受剪应力。

例如，如果一个建筑师想在简支梁上开孔（通电线或水管），在梁跨中处没有问题，因为这里力很小，两侧翼缘可以承受弯曲荷载。但在支座处不应开孔，因为腹板要承受很大的剪荷载（图 29）。

注释：

电缆、上水管、污水管和通风管道可以对承重结构产生关键性的影响。它们应在早期安装并与结构工程师达成一致。最重要的是承重结构和管线要好好规划，以便尽可能地出现交叉点。

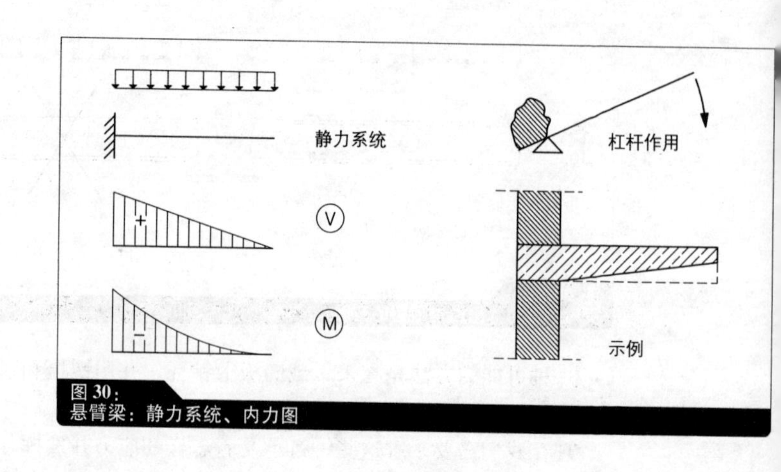

图 30：
悬臂梁：静力系统、内力图

P35

结构构件

P35

悬臂梁、 简支梁和带悬臂简支梁

前文以悬臂梁和简支梁为例说明了荷载和力。这两种承重系统是大多数复杂系统的基础。这里再简要阐述它们的优点和缺点。

悬臂梁

悬臂梁与用来提升重物的杠杆相似。在支点处的杠杆作用是最大的问题。如图 30 所示，锚固点是扭矩和剪力最大的点。悬臂梁很少在木结构中使用，因为除非锚固点很长，否则没有钉子或螺栓可以承担如此大的荷载。但在砌体结构中，尽管有如果锚固点上没有足够的重量，长的悬臂将把上面的砌体结构撬起来的风险，这样的锚固点还是可能的。我们观察弯矩和剪力图，很明显，承受均布荷载的悬臂梁尺寸取决于绕其锚固点的面积。但对其长度除锚固点以外的其他部分，尺寸是偏大的。所以为了节省材料，通常悬臂梁从其固定端到自由端，截面高度是逐渐减小的，就像弯矩图一样。

简支梁

简支梁可能是最普通的承重系统，需要仔细地进行研究。简支梁通常采用简单的均匀截面的木、钢或混凝土截面。因为它们生产简单且便宜，并且能够在底部和顶部提供平面。但通常简支梁只在一点上被充分利用，也就是跨中最大弯矩处。所以将简支梁按弯矩图的形状制造，跨中尺寸大于支座处尺寸是合理的。（图 31，见“内力”、“弯矩”）

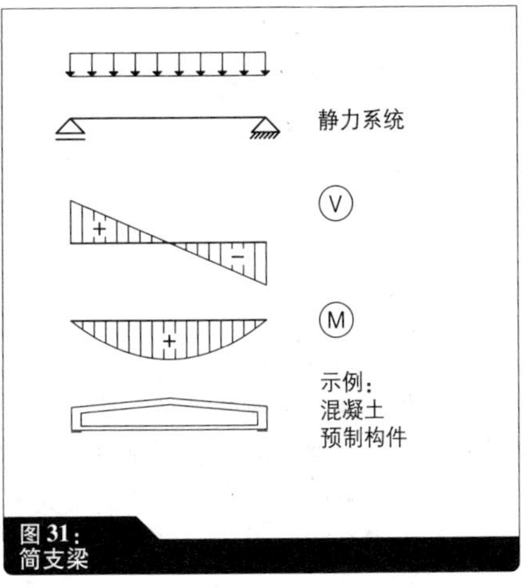

图 31：
简支梁

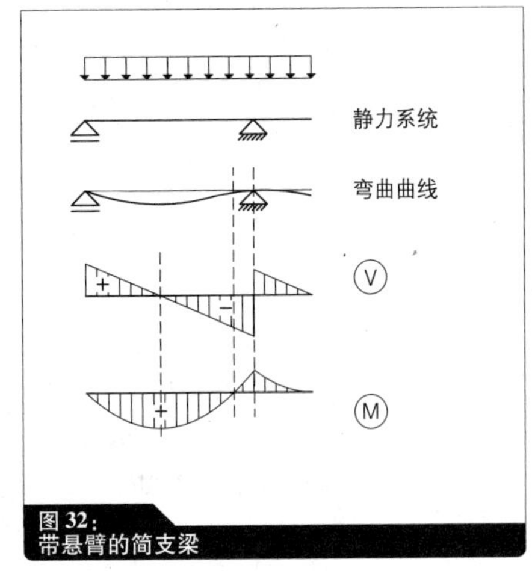

图 32：
带悬臂的简支梁

木材是一种天然材料，其沿纤维纵向比其横向能够承担更大的荷载，因此其对剪力是很敏感的。在最好的情况下，木梁在三点被充分利用，即跨中弯矩最大处和两个支座的剪力最大处（见“确定尺寸”，“剪应力”等节）。

带悬臂的简支梁

从承重理论讲，带悬臂的简支梁是一种非常有用的系统。可以说，作为上述两种系统的结合，它弥补了两种系统的缺陷。悬臂梁的问题在它的锚固点。但是在带悬臂的简支梁中，锚固点的长度就是梁的跨度，它往往大于悬臂的长度，因此锚固点长度也就不成问题。这种梁的关键在于悬臂梁的支座处发生了什么？在支座处，悬臂梁截面有最大的负弯矩。（图 32）

简支梁在跨中有最大正弯矩，而在支撑处为零。这两个弯矩图（简支梁和悬臂梁）如何结合在一起？如果我们想像这种构件在加载条件下变形就非常清楚了。悬臂梁向下，构件在梁跨处也将向下弯曲。在一端铰支座处向下，而在另一铰支座处，梁是水平的。

弯曲线

这意味着弯曲线的拐点从支座处移到了梁跨内。

这也可以从弯矩图中反映出来。与突出部分相对应，负弯矩最大值在支座处。

支座弯矩

跨中弯矩

P37

连续梁效果

P38

零弯矩点

支座处的负弯矩称为支座弯矩。它将一直在梁跨内延续，直到“跨中弯矩”出现。这是一个两支座中间区域正弯矩的术语。因为支座处弯矩，跨中弯矩小于纯简支梁条件下的弯矩。跨上的梁由于悬臂梁的存在而尺寸减小了。这意味着在相同跨度时，这种梁的尺寸要小于简支梁。

连续梁

连续梁是跨越几跨的梁。它们根据跨数来准确定义。一根两跨的连续梁有 3 个支座，三跨有 4 个，以此类推。这种系统是上述系统的逻辑外延。在带悬臂的简支梁中，在中心支座处会产生支座弯矩，同时减小了跨中弯矩。在这里，尽管弯曲变形曲线与弯矩曲线的形式不同，变形曲线的弯曲点与弯矩曲线的零点是相对应的。但弯曲变形曲线的形式可以表现出弯矩图（图 33，“内力”，“弯矩”等节）。

因此连续构件的优点在于可以通过支座弯矩减小跨中弯矩。跨中弯矩的减小意味着梁截面的减小。

连续梁能够大大节省材料。

如果只是简单地想一下，我们似乎可以看出中心支座承受的荷载两倍于边支座。而事实却并非如此。荷载源区的起点并不是在跨中，而是在剪力为零而跨内弯矩最大处。因此中心支座承受的荷载比边支座的两倍要大（图 34）。

铰接梁

当观察连续梁的弯矩图时，会发现另一种可能性：在零弯矩点处可将单根梁组合起来。这保持了梁的连续性，弯矩为零的点意味着该点没有弯曲效应。所以如果在这点上设计一个梁节点，与连续梁相

注释：

梁的方形截面既不能太宽也不能太窄。高宽比在 2/3 和 1/3 之间比较合理。建筑图表给出了木截面的尺寸。表中列出的各种尺寸的截面都可以在市场上获得，因此它们不必加工到特定尺寸，这减少了木匠的工作量。

提示：

正弯矩或跨中弯矩以截面下边缘受拉、上边缘受压为特征。负弯矩或支座弯矩截面上边缘受拉、下边缘受压（“内力”，“弯矩”等节）。

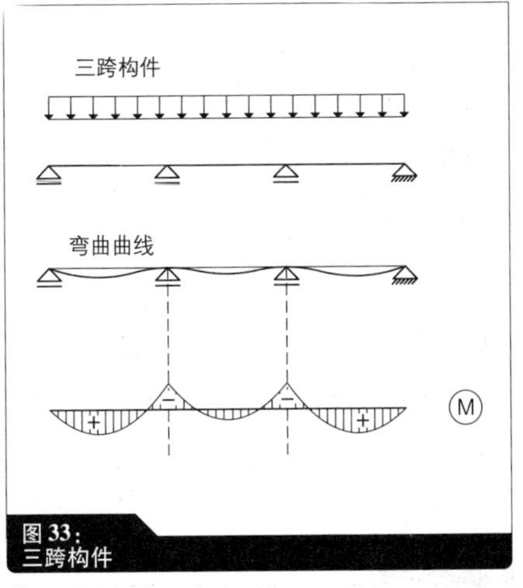

图 33：
三跨构件

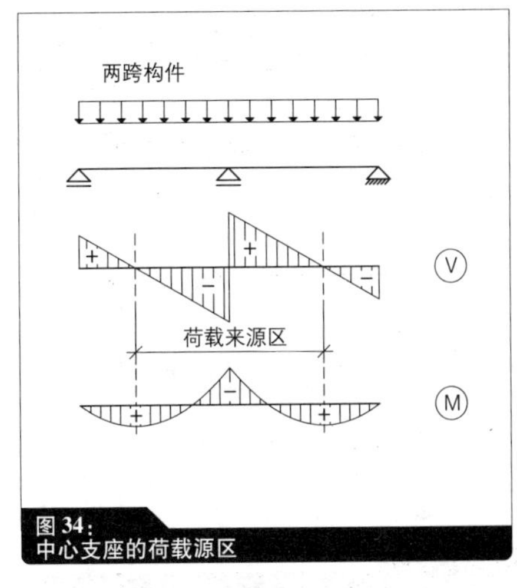

图 34：
中心支座的荷载源区

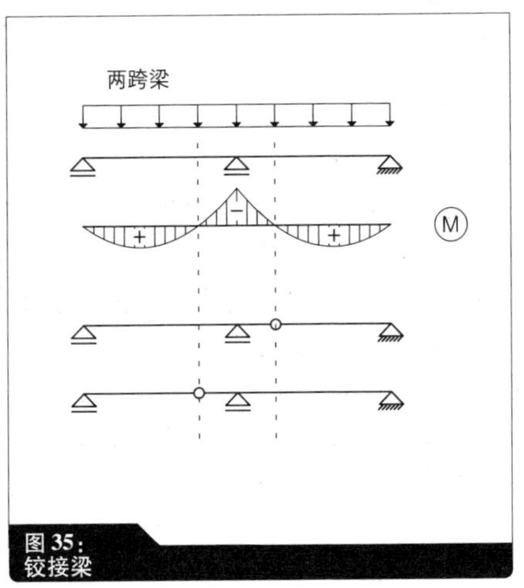

图 35：
铰接梁

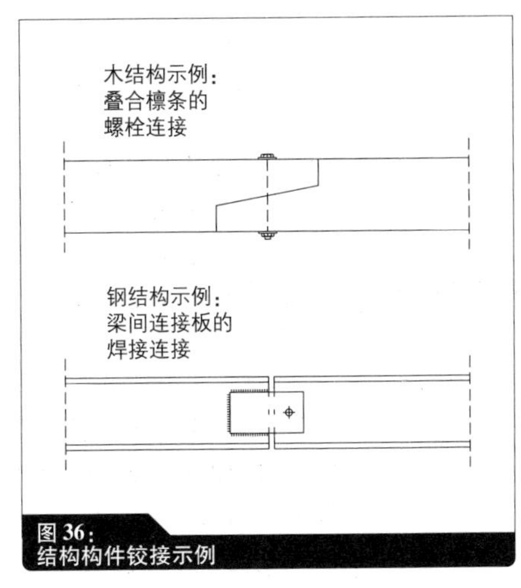

图 36：
结构构件铰接示例

比，弯矩图不会发生改变。在木结构中，几乎在每个节点都用到铰接的梁节点。在铰节点处弯矩必然为零（图 35，图 36）。

静力测定

增加的铰节点可以进一步地影响整个系统。连续梁和铰接梁在本质上是不同的。如果一个支座由于某些原因下降，连续梁会发生什么

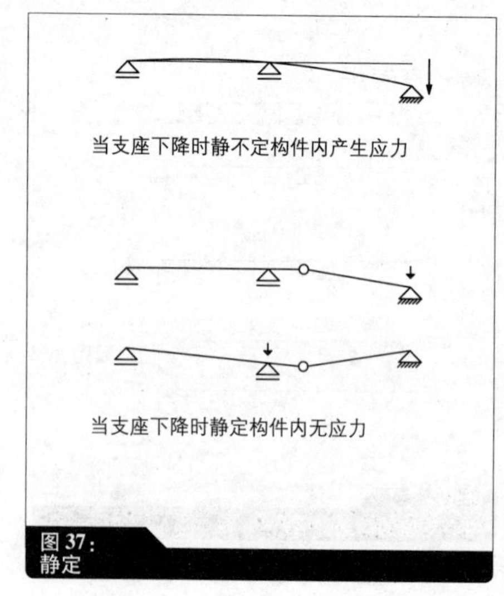

图 37：
静定

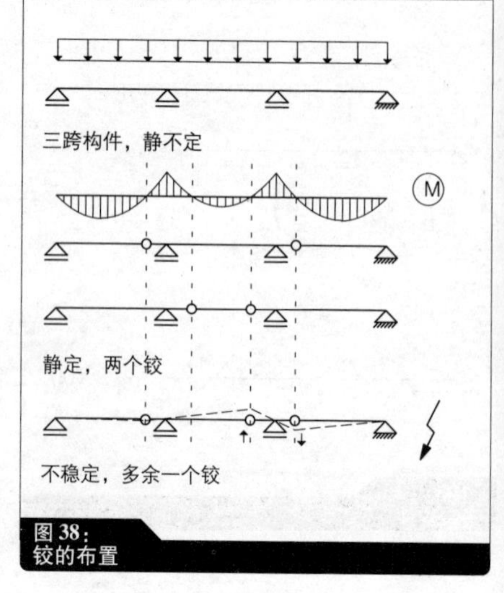

图 38：
铰的布置

呢？为了保证所有的支座可以发挥作用，梁不得不产生弯曲。这在结构构件内产生应力。如果铰接梁发生这种情况，截面内将不会产生应力，因为支撑系统是铰接且可以转动的。

静不定系统

如果移走一个支座，承重系统产生内力，则该系统叫做静不定系统（超静定系统）。如果没有产生内力，则称为静定系统（图 37）。

静定系统

例如，悬臂梁和简支梁是静定系统。下一章中将提出更多的静定和静不定承重系统。静定主要依靠支座的数量和性质，以及铰节点的数量。增加铰节点可以使静不定系统转变为静定系统。但要小心，过多的节点将使结构趋于不稳定。

三跨构件

一个统一分配荷载的三跨构件表明它需要两个铰节点来升高或降低任一节点，而不产生内力。换句话说，需要两个铰节点来生成静定系统。力矩分析中显示了四个零力矩点，因此如何安排这些节点有好几种可能性（图 38）。

在实际中，静定和静不定之间有什么区别呢？静不定系统可以提供更大的安全系数。例如，如果连续梁的一个支座失效，结构构件可能不会发生倒塌，因为构件仍然由剩余的支座支撑。在静定系统中，例如简支梁，却不是这样。另外，静不定系统无法利用三个平衡方程计算。在这里需要更多更复杂的计算方法。

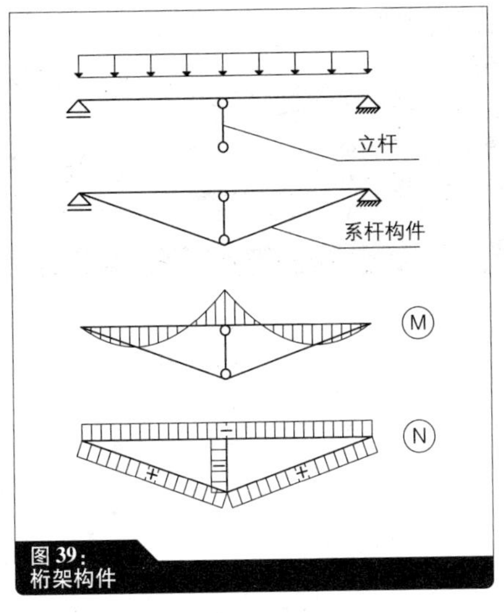

图 39：
桁架构件

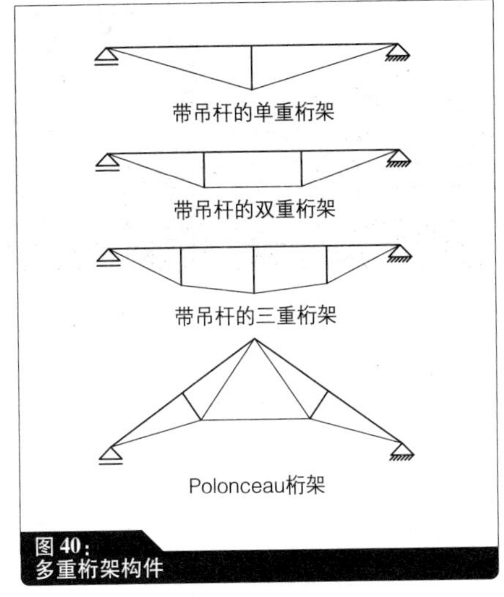

图 40：
多重桁架构件

P40

桁架梁

跨长是选择承重系统的最重要的标准。在任何结构中，都可以确定一个跨度的临界点，在临界点内梁跨可以发挥功能，而超出后将不再有效。例如，对简支的木梁来说，这个临界点大概是 5 ~ 6 米。如果跨度加大，则需要进一步的措施：例如，如果下面无法放置一个支撑，那么可以插入一个支架（图 39），可以桁架的形式承担荷载。尽管没有接触到地面，桁架类似一把张开的弓，将支架向上推，发挥像支座一样的作用。这种系统叫做桁架构件或桁架梁。

也可以利用双重或三重桁架支撑梁（图 40）。尽管相应的结构构件内的力相应增加，但跨度可以进一步加大。桁架构件的单个部件是如何受力的呢？当支撑梁时，支架承受压力；桁架（通常由钢杆制成）承受拉力。梁原来只是承受弯曲荷载，现在产生压缩来抵抗拉力。

Jean B. -C. Polonceau (1813 – 1859)

用撑杆和桁架我们可以构造出更多复杂的系统。一个例子就是 Polonceau 桁架（以它的发明者命名，图 40）。

可以总结一下桁架构件的特点：简单的梁成为复杂的系统，不但承受弯曲荷载，并且在相距一定距离的不同结构构件内利用压缩和拉伸来承担荷载。上面的构件不单承担弯曲荷载，同时要承担压力，而桁架承担拉力。当解释弯矩时，我们提到了受压截面的力臂。这个力

图 41：
钢格构结构

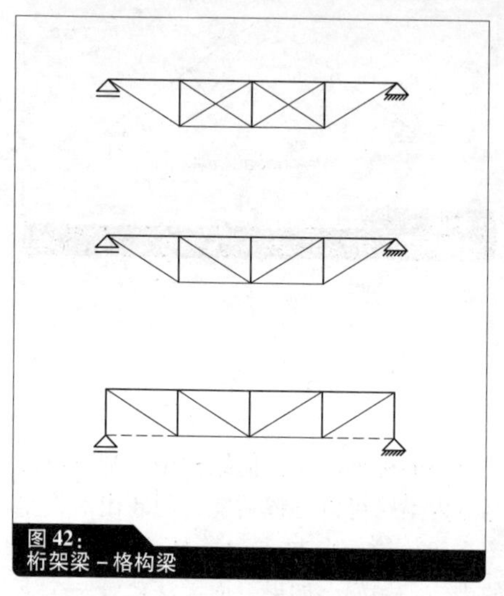
图 42：
桁架梁 – 格构梁

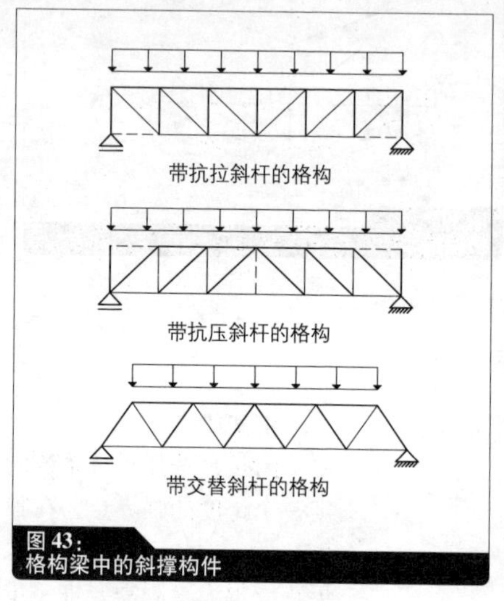

图 43：
格构梁中的斜撑构件

臂在这里很明显被放大了。因此这种系统更为高效（见“尺寸”、“弯矩抵抗力”节）。

P41

格构

具有三个以上支架的桁架梁在许多方面是不合理的。但是如果支架的各个截面单独被支撑，这就产生了一种新的可以具有更大跨度的系统——格构或骨架梁。在这些格构梁中，受拉截面通常不是由索或杆，而是由木或钢构件组成。格构是常见的有效系统，能够不断被改造使其满足特定环境的需要。它们可以利用任何材料建造，且杆可以用截然不同的形式布置（图 42）。

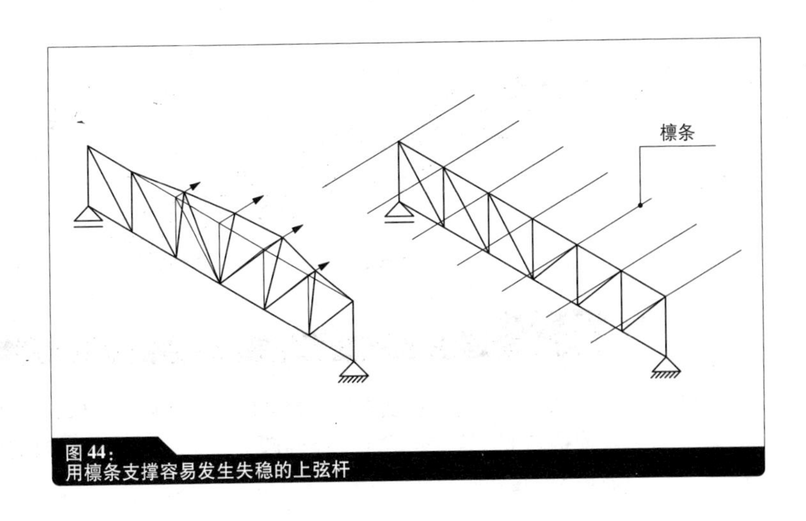

图 44：
用檩条支撑容易发生失稳的上弦杆

受拉斜撑

在目前讨论的例子中，与下面的桁架相对应，斜撑可以是受拉杆件。但它们也可以按相反的方式布置。

受压斜撑

于是它们将承受压力。为确定斜撑内力的方向，考虑它们是承受压力还是如同下垂的索一样承受拉力是有帮助的。

可变换斜撑

也可以按不同方向只布置斜撑来构造格构桁架。在中心处，桁架的外形并未改变。一个方向的杆件受压而另一个方向受拉，在格构结构的中心处，构件它们承受的荷载发生了改变，而布置却并未发生变化。

无力杆件

在格构梁内有不直接承载的杆件。它们既未受拉也未受压，叫做无力杆件。但是因为结构的原因，通常它们是不能省略的。这意味着它们组成了系统的形状或者说是将系统保持在现有位置。在图中，受压构件用粗线表示，受拉杆件用细线表示，无力杆件用虚线表示（图43）。

格构梁的高度和长度按照跨度计算。但是宽度只依赖于各种情况下选择的梁截面，与全长相比通常是很窄的。为此，上部的受压截面（叫做上弦杆）很容易产生失稳（图 44）。

上弦杆

这个问题可以采用不同的方法解决。

檩条

上弦杆可以固定在上面的顶棚或纵向顶棚梁上，以防止其移位，或者本身就可以制成防失稳梁。

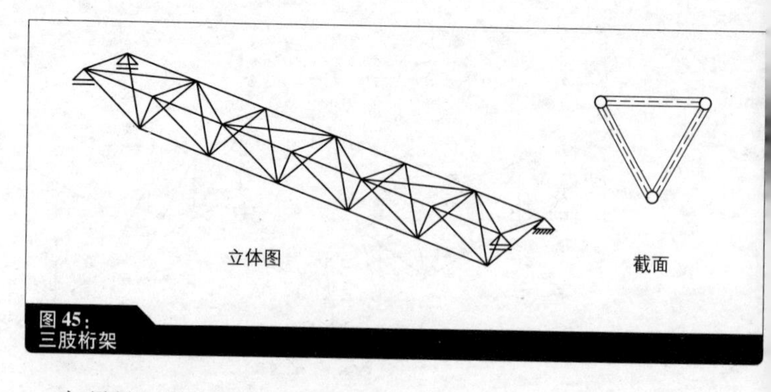

图 45：
三肢桁架

三肢桁架

如果另一根弦杆加在有失稳危险的上弦杆上，然后二者之间用斜撑连接起来，这就形成了一个在两个方向上的刚性支撑构件，叫做三肢桁架（图 45）。

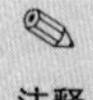

P44

板

木或钢结构几乎都是方向性系统，也就是说，杆型的构件意味着只能在一个特定方向上承担荷载。然而，混凝土却可以构造无方向性的扁平构件。

钢筋混凝土

基本上，钢筋混凝土的力学性质如下：作为由水泥、水和骨料（如砾石和碎石）形成的人造石，混凝土承压性能较好，但却如同砖石一样无法承受较大的拉力。因此，混凝土总是和钢筋组合使用。

钢筋

在这种组合材料结构中，混凝土承受压力，钢筋承受拉力。前面关于不同支撑形式的一章中已经解释了在结构构件内何处存在拉力。这就是钢筋混凝土内钢筋摆放的正确位置。对于板来说，需要在下部

注释：

格构柱的节点应使截面或更精确地说，是截面的中心线精确地汇交到同一点。这样能避免节点产生扭转力，从而对节点产生附加荷载。

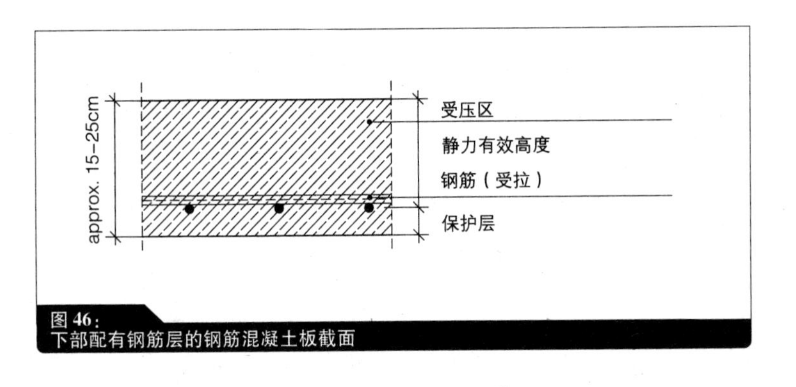

图 46：
下部配有钢筋层的钢筋混凝土板截面

和周边区域设置钢筋。如果钢筋混凝土板像连续梁一样使用，钢筋设置在上部。当浇筑楼板时，钢筋通常以交叉钢筋网的形式布置，在各个方向被混凝土包围，从而能够承担复合荷载。混凝土楼板厚度取决于跨度，通常为 15～25 厘米（图 46）。

混凝土板几乎是惟一无方向性的构件。在一个四方区域，混凝土板可以同时向四面的墙传递荷载。但如果板是矩形的，荷载将首先通过短边承担，因为如果变形相同，短边将比长边承担更大的荷载，由此产生更大的应力。对于长短边比为 2 的板，长边几乎不承受荷载。但是钢筋并不是根据施加荷载的主方向布置。板总是沿横向布置钢筋，这是平面效应的优点。例如，这意味着集中荷载可以被更好地分散，从而使板内的力减小。

跨度越大，需要的板厚越大。但如果板的厚度超过 25 厘米，其自身的恒荷载将变得非常大，以至于实心平板将很难应用。严格地讲，只有板的上缘能够有效地承受压力，钢筋承受拉力。结构的其他部分实际上只是连接或填充的作用。

肋板

如果板很厚，通过减少下部区域的面积以减轻自重是非常合理的。钢筋主要布置在板肋内，彼此之间非常紧密。肋板通常比平板跨度更大。

铰接点

跨越更大跨度的另一种方法是使用连接格栅。与肋不同，连接格栅不被认为是板面积的一部分，而是当作上面放有平板的梁（图 47 和 132 页图 69）。

板梁

对于在施工现场浇筑的钢筋混凝土结构（现浇混凝土结构），连

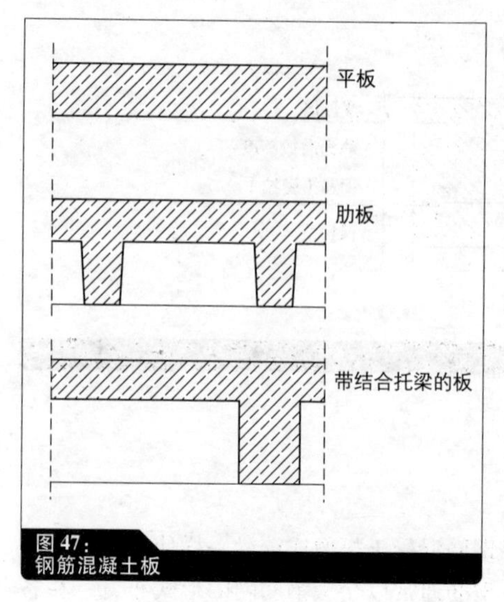

图 47：
钢筋混凝土板

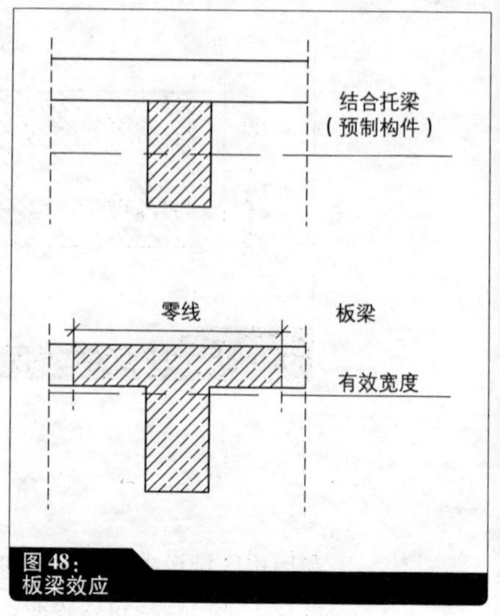

图 48：
板梁效应

接格栅最好利用巨型施工方法。这里，“巨型”的意思是所有的，甚至是不同阶段浇筑的现场混凝土构件，都以连续结构的方式工作。因此连接格栅不仅利用了肋到板下边缘的静态高度，同样也利用了板的厚度。另外，梁在各个方向支撑着板，扩大了受压区域。在这种情况下，可以用板梁这个词。（图 48）

P46

柱

与水平承载构件不同，柱基本上不承受弯曲荷载，而主要是承受轴向荷载。如果不考虑柱可能侧向弯曲而失效，很小的柱截面就可以承受较大的轴力。

失稳

细长柱有失稳的危险。但这种危险性取决于很多因素。柱的特征参数有荷载、材料以及长细比。瑞士数学家莱昂纳多·欧拉（Leonard Euler，1707－1783）确定了柱上下两端连接方式对失稳性能的影响，并将其划分为以他的名字命名的四种情况。

欧拉准则

欧拉准则提出了以支座或铰进行连接的柱的 4 种情况。当失稳时，柱的变形曲线满足正弦曲线形式。柱的边界条件影响正弦曲线的长度或拐点的位置。这对柱的稳定性有着重要的影响。与变形曲线相关的柱长称为有效或失稳长度。

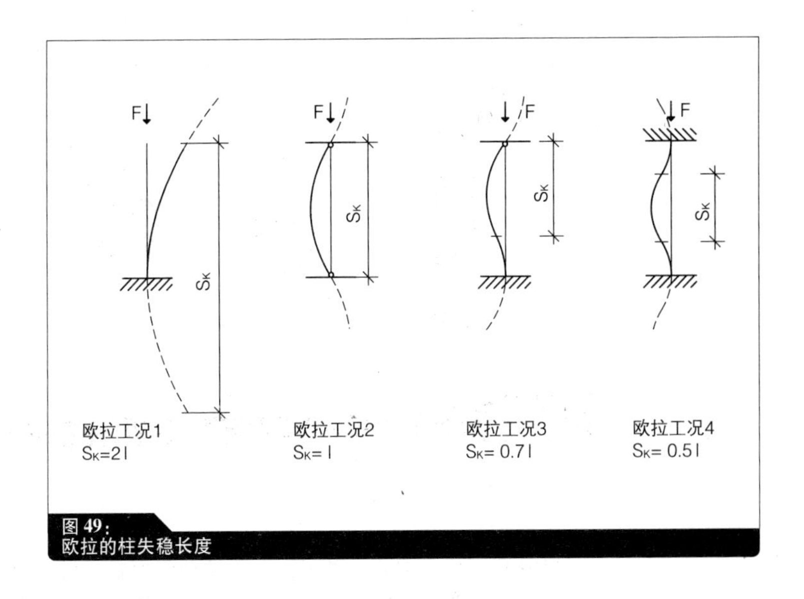

图 49：
欧拉的柱失稳长度

图 49 显示了具有相同柱长的四种情况。第一种是一端固接一段自由的情况（类似于旗杆）：变形曲线很长，稳定性不好。第二种情况是柱的上下两端皆为铰接。这种情况很常见，变形曲线或失稳长度较第一种短，使得柱更加稳定。第三种是在第二种的基础上将一端固接，从而限制了该端的转动，减小了正弦曲线的长度，即失稳长度。第四种是上下两端皆为固接，柱的失稳长度最短，因而柱也最稳定。

长细比

决定柱的尺寸的另一个主要参数是长细比。可能会认为长细比是柱长度与厚度的比值，但实际情况却不是这样。厚度并不包括在等式内，而截面惯性矩与面积的比值因为表明了稳定性，从而成为等式的一部分。而上面提到的欧拉失稳长度（而不是柱长）是非常重要的。柱的长细比就是失稳长度与弯曲强度的比值。

提示：

柱的欧拉失稳性能是以假设材料具备拉压性能为前提，不适用于确定圬工或混凝土柱的尺寸。

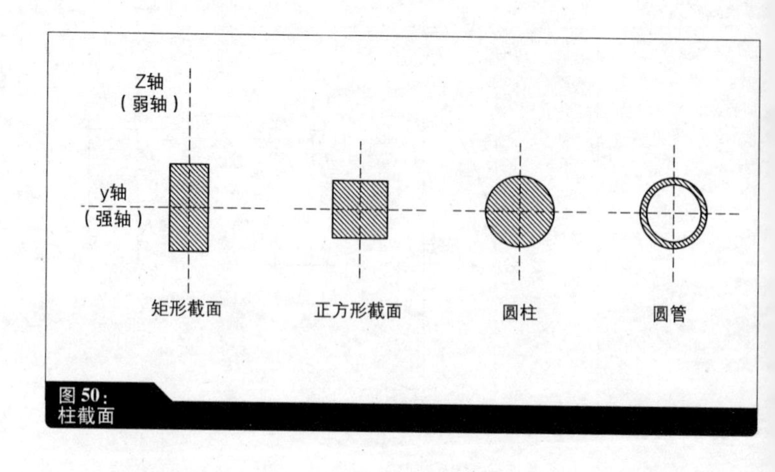

图 50：
柱截面

可以利用这些参数来确定柱的最优截面形状。只承受竖向荷载的柱有可能向任意方向失稳。然而，实际上，柱通常向最小刚度方向失稳。所以柱应该在每个方向上都同样稳定。柱截面应采用矩形，如果采用圆形截面会更佳。

另外，基于与惯性矩相关的弯曲强度，可以进一步得到关于理想截面的一些结论。根据失稳柱内的应力分布，很明显距零应力平面或中心较远的区域是最有效的。在管状截面中，去掉了起不到太大作用的中心区域，这样材料距中心能尽可能远地布置。这说明，管，特别是圆管，对柱来说是最好的截面形式。这个结论只是理论上的，用来说明最终当很多其他因素影响结构时，柱的承受荷载情况。所有这些在设计中都应被考虑到（图 50，图 51）

P49

索

索的工作原理与前面各章所讲述的原理都不相同。如果索是承载结构的一部分，它将因为上面作用的荷载和自身重量而下垂，并随荷载的变化改变形状。索无法承受弯曲，总是保持没有弯矩的形状。这种形状与梁，而不是索的弯矩图精确一致。

索线

所以“索线”是与其曲率相对应的（图 52）。

索结构与前面讨论过的承重结构的另一个不同之处是在支座处总是有水平反力作用。索以轴力，即索拉力的形式承担荷载以及水平反力，索拉力方向在支座处与索的方向是一致的。只有当索竖直悬挂

图 51：
柱

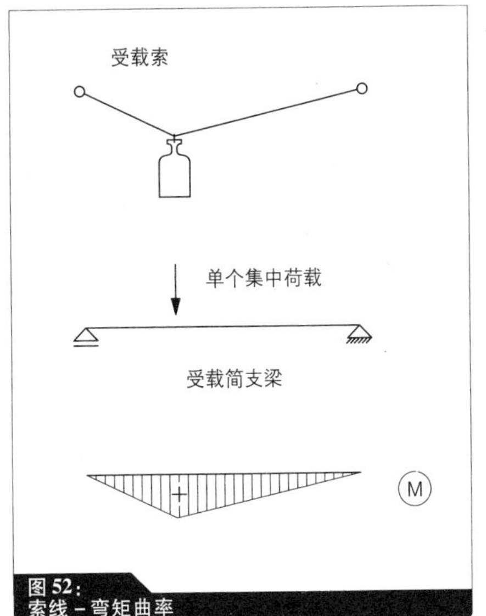

图 52：
索线 – 弯矩曲率

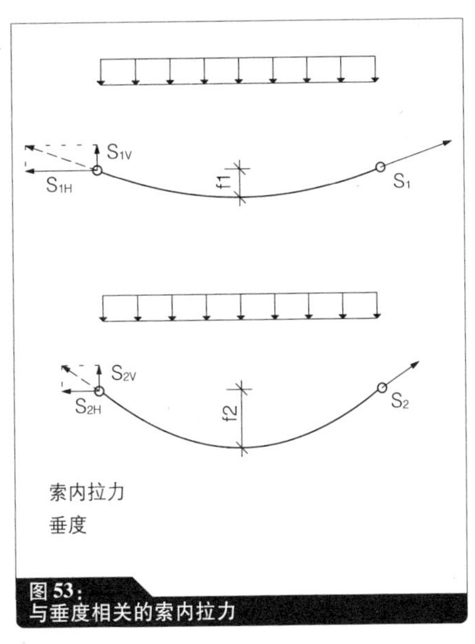

图 53：
与垂度相关的索内拉力

垂度

时，在支座处才只有竖向反力（96 页图 13）。在图 53 中，比较两种索可以发现所得竖向分力与承受的荷载幅值大小相等，而水平分力随索角度的改变，即松弛程度的改变而变化。

索力

我们可以在张紧或放松的晾衣绳中体会到索结构的重要参数：当垂度较小时，索力较大，而垂度较大时，索力反而很小。

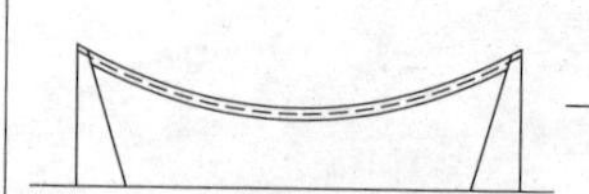

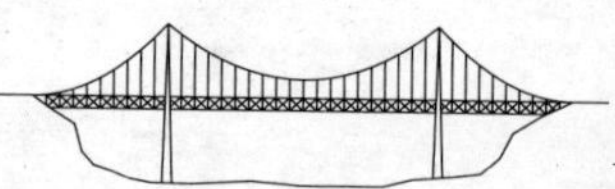

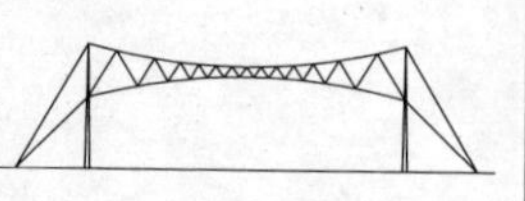

图 54：
悬索结构的加劲

那么为什么索结构不能更普遍地应用呢？索结构在实际中应用时未知因素较多。它们的允许位移值较小，给施工带来了较大的难度。一些无法控制的移动，如风振效应，一定要减小到允许范围内，以免产生很大的动应力。缆索结构在各种情况下都必须保持稳定，但能做到这点的方法很少。一种可能的方法是增加缆索结构的自重，使得荷载变化或风荷载与自重比起来很小。例如，这种方法对于悬吊屋面是有效的。

但不利之处在于附加荷载通常以预制混凝土构件的形式施加，这意味着失去了索结构自身的优势，并且缆索需要承受较大的内力。

另一种方法是采用抗弯刚性结构构件加劲。例如，在悬索桥中，所悬吊的行车道是刚性的，从而加强了整个结构的刚度。

缆索承重结构也可以用其他缆索反向拉伸的办法来加劲。这可以用不同的方式实现。可以利用二维梁，例如 Jawerth 桁架（这种结构的样式很容易让人想起格构结构，但实质上与其完全不同）。这种结构内所有的索都被张紧，即使在很大的荷载作用下也不会松弛。这意味着结构在形式上可以保持稳定并承受相应的荷载（图 54）。在由索网构成的二维构件中，系统刚度是通过向曲线反方向施加预应力形成的。（见“板结构”）

提示：

缆索承重结构中的索用高强钢材制成。许多细的钢丝（直径随索形式不同而不同）相互缠绕形成一股钢绞线。然后由多股钢绞线制成缆索。

图 55：
拱形承载系统

P51 ## 拱

如果将索的形状固定并翻转过来，我们就会得到另一种以压力而不是拉力承担荷载的结构形式。因为只承受轴力，因此如同索一样，这是理想的拱形式。

抗力线

这种理想的结构形式，可以用计算或画图的方法得到，叫做抗力线。

拱和索还有着其他共同之处。

拱高

拱在支座处也承担着竖向及水平荷载。如同索一样，水平反力大小也与拱的高度（自楼板或拱脚到拱顶的距离）有关。拱高越小，压力中水平分力的比重越大，这种水平分力叫做拱脚推力（图 56）。

索与拱最主要的不同在于实体拱与索不同，不能随荷载改变而改变形状。一条抗力线（作为一种精确的拱形式）只能用于一种单独的荷载位置。如果荷载改变，抗力线也随之改变。这意味着会在拱内同时产生轴力和弯矩。拱结构可以采用不同的方法来解决这些问题。

圬工拱桥的自重很大，作用在其上的荷载与自重相比很小，即使荷载改变，抗力线也几乎不发生变化，拱仍然保持稳定。拱也可以用附加的结构构件加劲。例如，如果用石材在拱的周边砌起墙，就能够防止拱变形或丧失承载力。也可以用刚性材料，如分层木板或钢板建造拱。这里拱支座的高度要足够大以承受弯矩与轴力（图 57）。

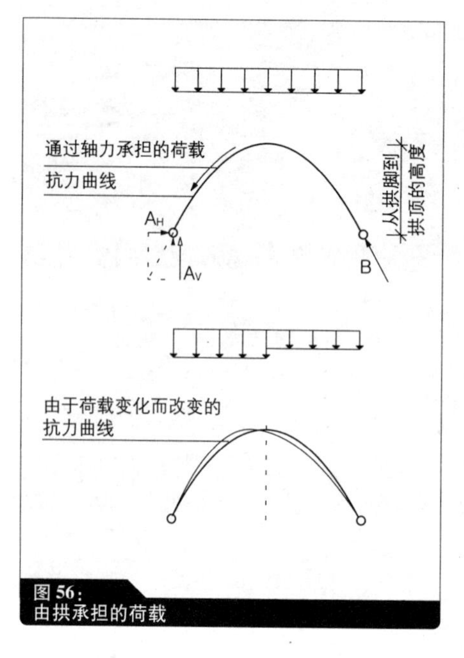

图 56：
由拱承担的荷载

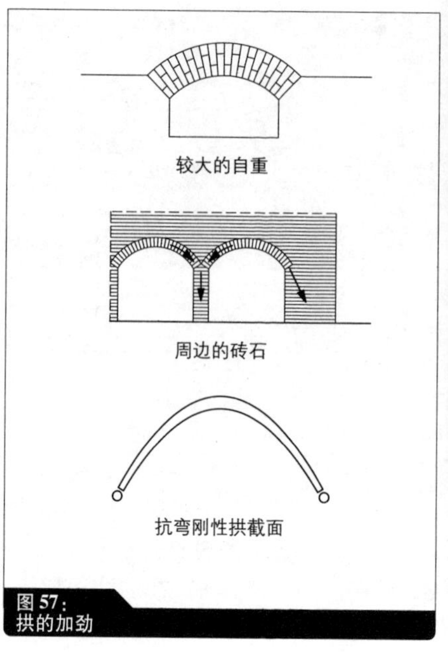

图 57：
拱的加劲

常用的拱一般有三种不同的静力系统：两铰拱、三铰拱和无铰拱。

两铰拱

两铰拱具有两个铰支座。它们承受竖向和水平反力，但不承受弯矩。想一下如果支座下移会发生什么？这个问题表明这是个静不定系统。

注释：

不要将拱形承载系统与拱形抗弯梁混淆。支座不承受水平力的拱只能通过弯曲的方式承受荷载作用。

提示：

拱形承载结构是从石建筑中演化而来的。因为石材只能承受轴力，所以所有的房屋开口（如门窗）都必须由拱支撑。老式的石结构为研究多种拱承载结构和拱推力的处理方式提供了很好的机会。

关于拱的更多的资料，可参见本套基础教材中的《砌体结构》（征订号：18859），中国建筑工业出版社 2010 年出版。

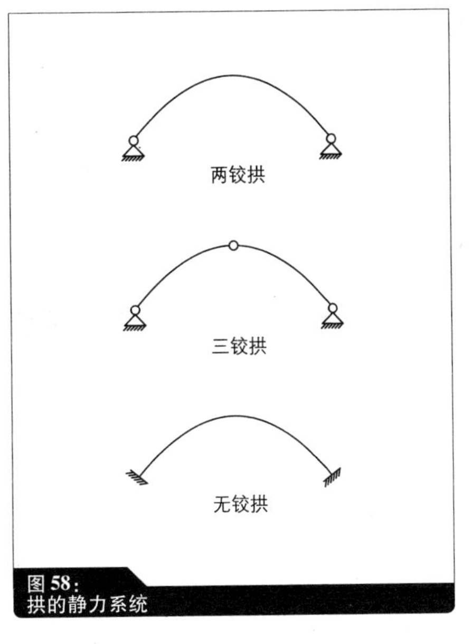

图 58：
拱的静力系统

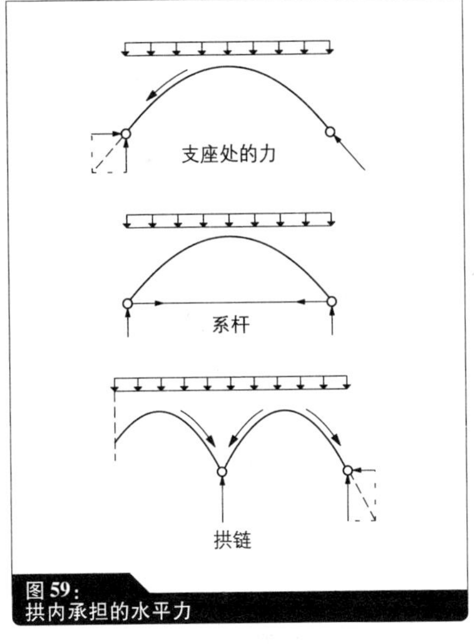

图 59：
拱内承担的水平力

三铰拱

再加上一个铰节点（通常是在拱顶），这个静不定系统就变成了静定系统。这几乎不会引起承载力的变化。但对于施工来说，它的优点在于可将一片拱拆成两部分运送，这样比较容易。然后两片拱可以方便地连接起来。

无铰拱（两端固接的拱）

将铰支座变成固定端可以使拱的刚度加大，因为固定端阻止了弯矩引起的弯曲。这种效应与柱欧拉失稳情况 2 和 4 一样。两端固接的拱是静不定的。这种拱通常不大，因为有效的固接施工非常复杂。（图 58）

拱的推力

有不同的方法处理产生的水平力，可以利用支座承受拱的推力，或在两支座中间设拉杆平衡两支座的水平力。如果几片拱连接在一起，则作用在支座上的水平力可以互相抵消，这时支座只承受竖向力（图 59）。

P54

刚架

设想有一个简单的承重系统，由两根柱以及上面的一根梁或桁架构成。如果柱的顶端和底部都为铰接时，系统不是稳定的。这时可以将水平的梁与柱用抗弯刚性节点连接起来，从而保证系统的稳定。这样就形成了一个有效的系统：刚架。

图 60：
钢刚架的交角

横梁

支杆

在刚架中水平构件叫做横梁，竖向支撑杆件叫做支杆。

当横梁与支杆刚性连接时，梁仿佛是“绕着梁和支杆的交角”转动。所以如果横梁在弯矩作用下弯曲，它可以将弯曲荷载传递到支杆。如果支杆没有横向支撑，则会向外变形。如果有支撑，就会限制这种变形，此时应力由刚架整体承担。支杆同时也限制了横梁向下弯曲。所以横梁并不是像简支梁一样工作，而是近似于两端固接的情况。

这也可以从弯矩图中明显地看出。由刚架交角的约束效应产生弯矩是刚架的特征之一。它的出现减小了刚架横梁的跨中弯矩，这意味着梁的尺寸可以适当减小（图 61）。

很明显，刚架的交角承受着由支撑弯矩所产生的荷载。为了获得足够的抗弯刚度，施工中应格外小心。由于简单的结构构件可以预制，因此分别制造横梁和支杆，并在施工现场将其组合连接起来是较为合理的。但这使获得抗弯刚性交角的问题更为严重，不过这种分别建造的方式也是刚架系统的一个主要优点。我们在开始时说过只有抗弯刚性交角使刚架成为稳定系统。刚性交角可以沿其纵向方向对刚架加劲，这对于框架系统是非常重要的，在静力系统中，刚架与剪力墙的作用是一样的，都用来加劲结构（图 61，见“加劲构件”一节）

双铰刚架

图 61 中的刚架，有两个铰支座，叫做双铰刚架。与双铰拱一样，是静不定系统。

三铰刚架

像拱一样，可以在刚架上再加一个铰节点，使其成为静定系统。这并不影响承载力，却对某些条件下的施工非常有利，特别是第三个铰可以按不同的方式布置，在跨中、边缘甚至刚架的交角处。由于铰节点处弯矩为零，所以该处的结构可比承受弯矩的区域纤细。

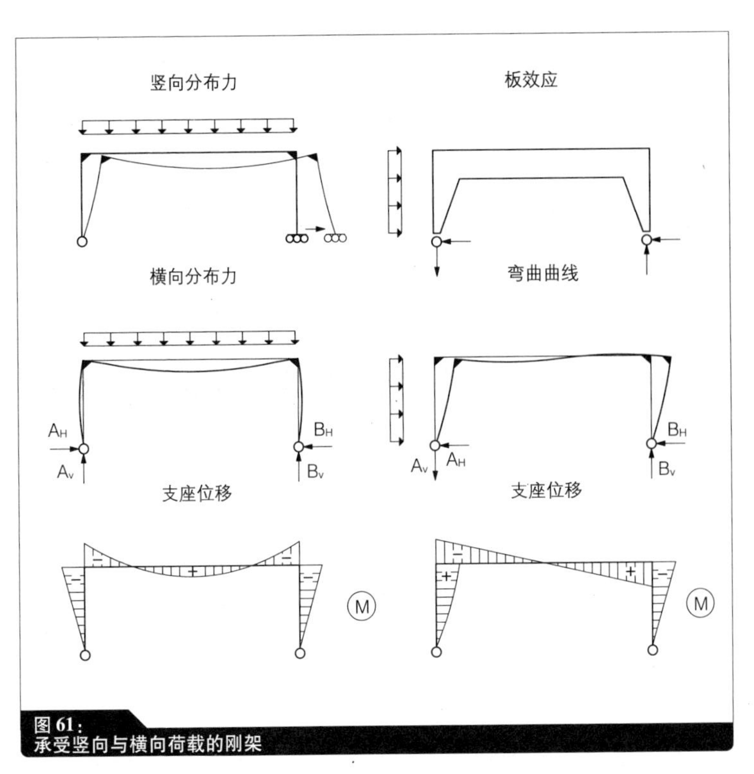

图 61：
承受竖向与横向荷载的刚架

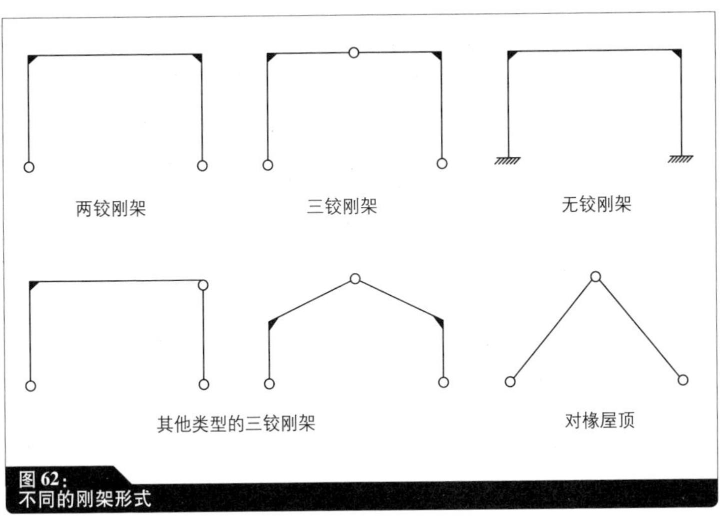

图 62：
不同的刚架形式

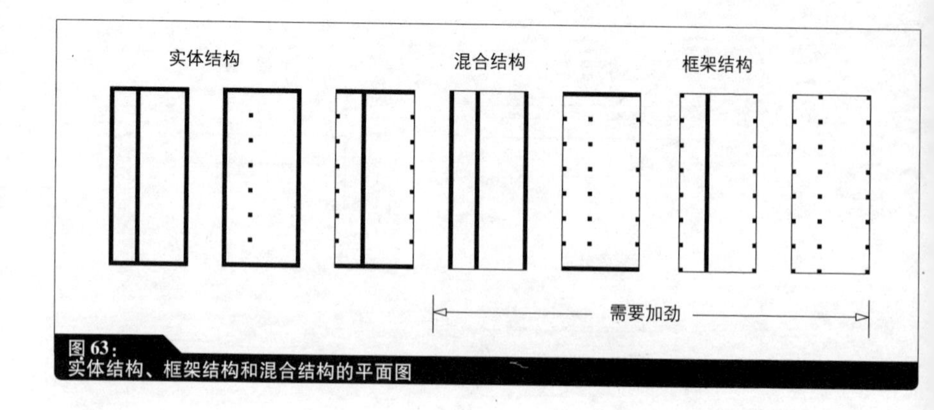

图 63：
实体结构、框架结构和混合结构的平面图

无铰刚架

如果在底部的支座为固接，可以进一步提高刚架的刚度。这时叫做无铰刚架。但这种结构形式很少使用，因为将支杆与地面固接是一个非常复杂的过程。（见图 62）

P59

承重结构

建筑是复杂的三维结构。它们的承重系统乍一看相当复杂、难以分析。但基本上所有的结构类型都可以划分为两种类别：实体结构和框架结构。这两种类别的结构在最初建造建筑时就已经应用了。目前所有的新技术都仍在这个范畴内。从古代的泥屋或吊脚楼到现代建筑的复杂系统，都应用这两种类别的结构。图 63 是实体结构、框架结构和混合结构的平面布置示例。

P59

板

实体结构

实体结构由承受水平和竖向荷载的平面构件组成。如同墙之类的板可以在自身纵向方向内承受水平和竖向荷载，但几乎无法承受横向荷载（图 64）。板或墙按不同方式失效，可能失稳或倒塌。当采用实体结构方法修建建筑时，可以利用其他墙的支撑防止墙板失效。支撑可按一定间隔或交叉布置。墙体互相支撑，使实体结构保持稳定。

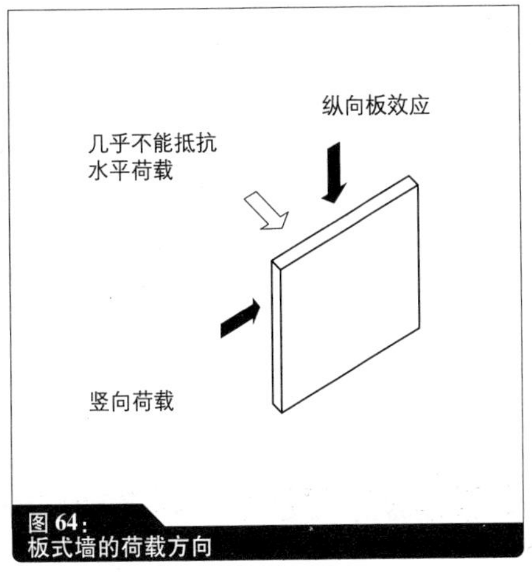

图 64：
板式墙的荷载方向

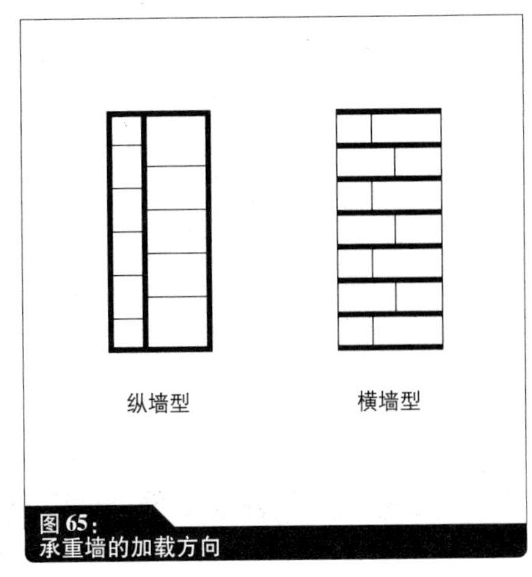

图 65：
承重墙的加载方向

模块结构

这种类型的结构也叫模块。我们可以将其分为承重式、支撑式和非承重式墙。非承重墙可以移走，并不影响结构的稳定性。支撑墙一般认为也是可以承重的。原则来讲，承重墙的厚度要大些，这使得它们可以承受楼板传下来的荷载。

实体结构可以分为纵墙型和横墙型。

纵墙型

如果一或两道主要承重墙与建筑长边平行，就是纵墙型，比较简单的城市住宅基本都属于这种类型。

横墙型

横墙型，又叫交叉墙结构，适合房间面积比较小的宾馆与排屋等。当使用带有只能单向承载木梁或预制混凝土构件的板时，这两种墙的类型可能会有所差别。但当使用可以多向承载的混凝土板时，纵墙与横墙通常都是承重的。(图 65)

砌体

最初的实体结构是砌体建筑。石墙不能承受拉应力，必须在高度、宽度和厚度方向加劲。在没有突起、支柱和开口的部位最好不要出现拉应变。

混凝土

“板”一章中曾经说过，钢筋混凝土也可以承受拉力，这意味着混凝土墙远比石墙稳定。根据房屋空间大小、跨度、开口情况和结构的复杂性，混凝土实体结构可以更自由地进行设计。它们可以现场浇筑或用预制构件进行建造。这些预制构件叫做“大板”，可以是小尺

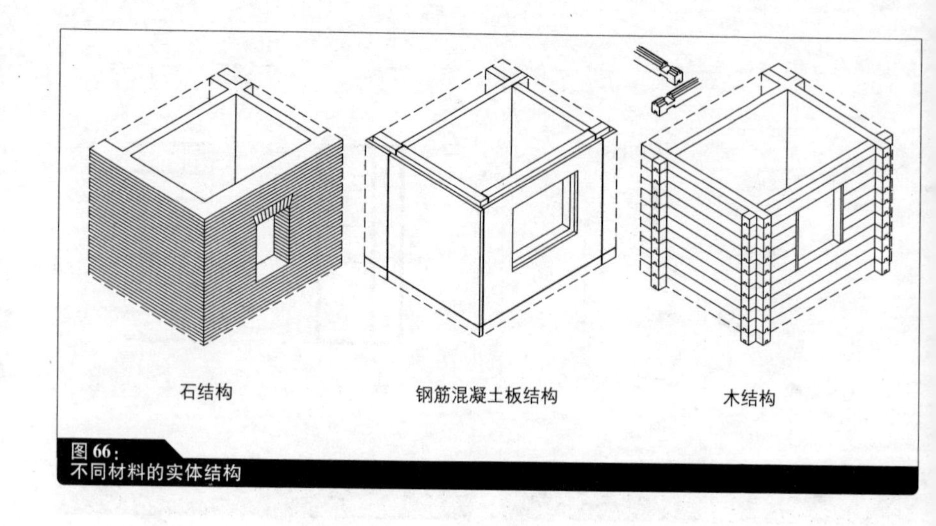

图 66：
不同材料的实体结构

寸的楼板或与房间尺寸相同的墙构件。

大板结构

采用大板建造建筑是目前流行的工业化建筑方法，叫做大板或板结构。各种构件用钢构件或混凝土连接起来，形成连续的、巨大的结构。

木材

尽管木结构通常采用框架法建造，其中有些还是应归入实体结构的类别。

木结构

首先就是木结构。木材水平垒放用作墙体。在屋角或建筑拐角处将木材连接起来从而使墙获得稳定。木材行业近年来进步很快，这几年市场上已有板材材料，这使得利用板材施工成为可能。其中一些是木板粘结在一起形成的板材，如同分层的木材；另一些是薄板交叉叠放形成的夹合板。这些板材材料的出现，使得现在的木结构施工方法与原有的一些方法有了很大不同，而这些方法目前还在进步。

提示：

这个名词“实体结构”（或“大块结构”）与“框架结构”对建筑师和对结构工程师来说，含义是不同的。上面的解释是基于几何学和结构学，以建筑学的语言给出的。对结构工程师来说，实体结构通常要涉及到砖石和钢筋混凝土材料。因此，结构工程师通常将该名词与材料联系在一起。

图 67：
框架结构

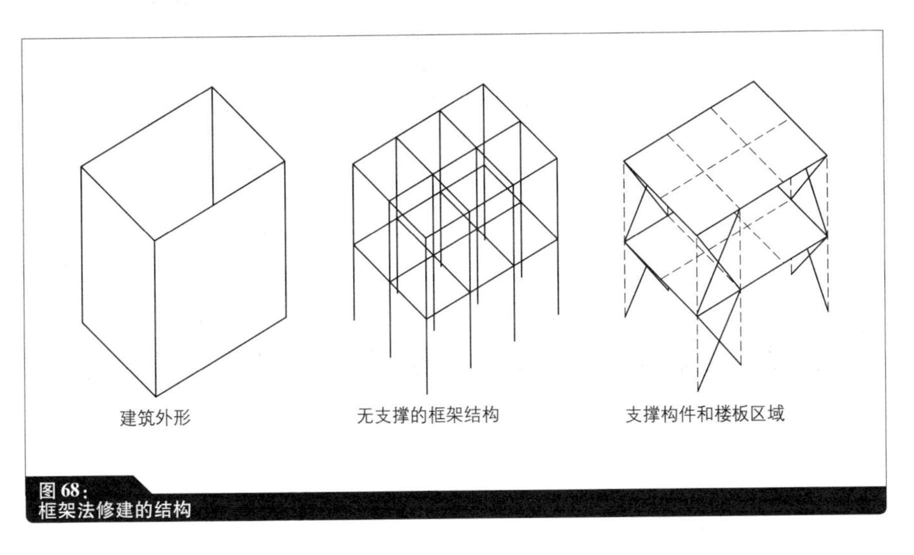

图 68：
框架法修建的结构

P62

框架结构

框架结构如同脚手架一样，由梁杆构件组成。然后板和墙构件在上面建造。承载构件和分割内部空间的构件，原则上是两个独立的系统（图 68）。

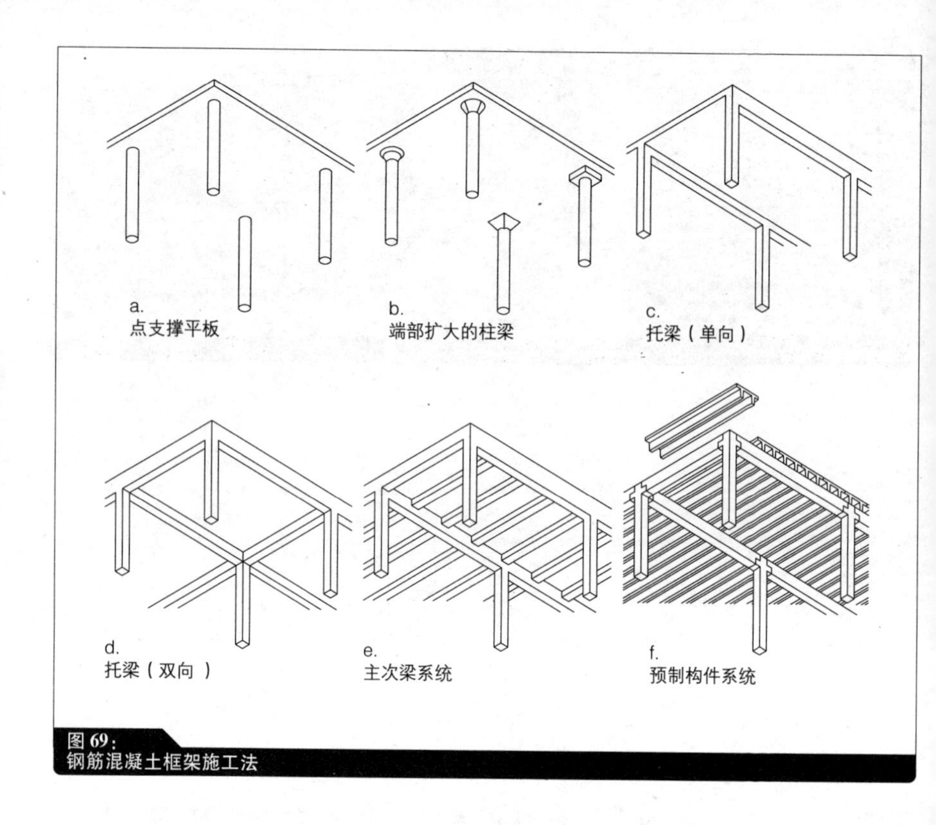

图 69：
钢筋混凝土框架施工法

框架结构基本上是由三类结构构件组成：柱、梁（包括板）、承担水平荷载的支撑结构。这些构件在节点处被恰当地组合在一起，通常是用铰连接。如果节点连接刚度不够，无法提供足够的约束，则连接就是铰接的。它们不必以铰链或类似的形式出现。原则上，所有能承受拉压作用的材料都可以用于框架结构。例如，木材、钢材或混凝土。每种材料都有自己的施工方法。同样在材料及其连接方法上也会产生一定的问题。

混凝土

可能框架结构最常用的材料就是钢筋混凝土了。无论是现浇混凝土还是预制钢筋混凝土构件都被广泛应用。通常是现浇实体钢筋混凝土板，然后建造支撑和柱。这种简易性说明了该系统的灵活性和经济性（图 69a）。但是板作为点支撑的平面结构，只能用于有限的跨度。所有板上的力都将传递给柱，这意味着柱和板的连接处将承受很大的荷载，柱有冲切入板的危险。

扩大柱端截面法

为避免这样的问题，板柱相交处的周边可以用不同的方式加强。其中一种是利用“扩大柱端截面法”（图69b）。

托梁

如果对于这种系统来说，跨度过大，那么就要用到托梁了。托梁横跨在柱间，沿着长度方向支撑着板。托梁可以用不同的方式布置。根据跨度不同，它们可以被设计为单向、双向或作为系统的主梁和次梁（图69c，d，e）。

框架结构也可以采用预制构件进行建造，有不同的包含顶棚、托梁、柱和基础的预制系统。这些构件尺寸大小是由如何能将其运送到施工现场所决定的，为了经济，理想的构件尺寸不应超过运输规定的尺寸。这些构件总是在现场组合并牢固地连接起来。这意味着从根本上来说，托梁是铰接的。

π形板

π形板（而不是平板），通常用作预制板构件。它们是窄的板面下带有两条肋，然后组合在一起形成楼板。它们如同T形梁一样工作，可以跨越较大的跨度。这种板放在梁上，而梁置于柱顶（图69f，“板”一节）。

钢

钢结构几乎都是框架结构，通常由不同外形的标准钢构件组成，叫做“轧制构件”（图70）。需要的尺寸由静力计算获得。

轧制构件

轧制构件最高为60厘米。如果需要更高的截面，那只能由钢板焊接而成（在钢结构中，几厘米厚的构件仍然叫做钢板）。

波纹钢板

波纹钢板通常用于覆盖大面积结构。梯形的褶皱使其具有足够的承载力和跨越能力，通常用于楼板或结构屋面（图71）。

钢构件在工厂中制成可运输的尺寸，然后在现场组合起来。焊接是最好和最简单的制作钢结构构件的方式，但当施工现场难以进行焊接操作时，通常用螺栓来进行组合连接。

提示：

板结构对楼层净高影响很大。因此，它们应该在设计初期就被考虑。板跨越宽，结构高度越大。

注释：

现有的结构参考书都有钢结构构件表，精确的给出构件尺寸和力学特征值。一般来说，在结构中应该采用这些构件，因为它们可以方便地从任何钢结构公司得到且十分经济。

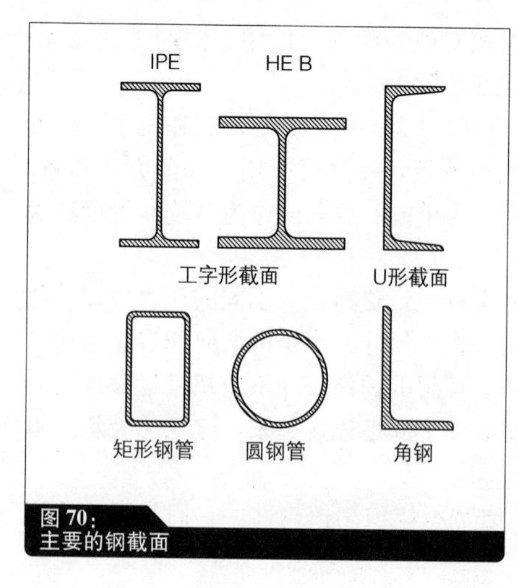

图 70：
主要的钢截面

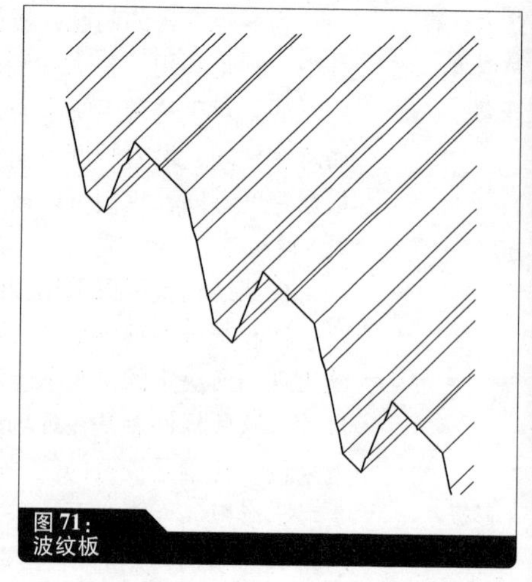
图 71：
波纹板

在钢结构中，可以利用一定人力物力制成抗弯刚性交角。这意味着柱和梁可以连接在一起形成刚架，发挥其承载能力（“刚架”）。为承担交角处较大的内力，这些点上的连接应该足够强。例如对工字形截面，横梁的翼缘应该用翼缘板与支杆连接，且连接螺栓应当尽可能地互相远离。相反地，对于铰节点，肋可以用简单的金属板连接件进行连接（图 72）。

防火

在火中，钢结构比木结构危险更大。在高温下，钢材软化，并迅速完全丧失承载力。所以在高层建筑中，必须避免钢材的火灾危险，如可以用石膏或泡沫涂料封装。

组合结构

钢结构与混凝土组合形成组合结构也可以降低其遇火升温的速率。例如在这些组合结构中，可以将混凝土填入钢管中，或工字形截面用混凝土填充。除减慢升温速率外，混凝土在火灾中还可以保持一定的残余承载力（图 73）。

木材

原木是最早的木结构材料。不同的文明都有一些古老而复杂的木结构技术。之所以复杂，是因为木结构建造方法的多样化和无穷多的变化组合。这里仅列出一些最重要的分类。

传统的木结构

传统的木刚架结构是纯正的框架结构并以泥或砖填充。作为一种手工建筑方法，它以精巧的连接而没有任何金属连接件而著称。木刚架结构现在很少修建，一般都是在历史遗迹中发现它们的踪迹。

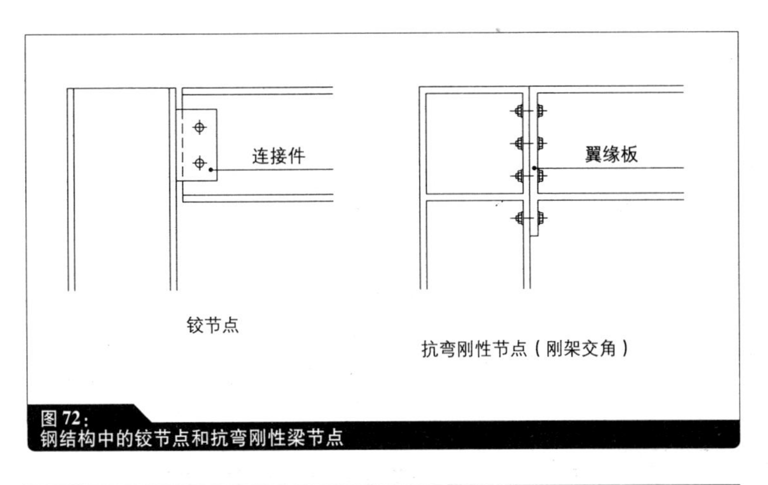

图 72：
钢结构中的铰节点和抗弯刚性梁节点

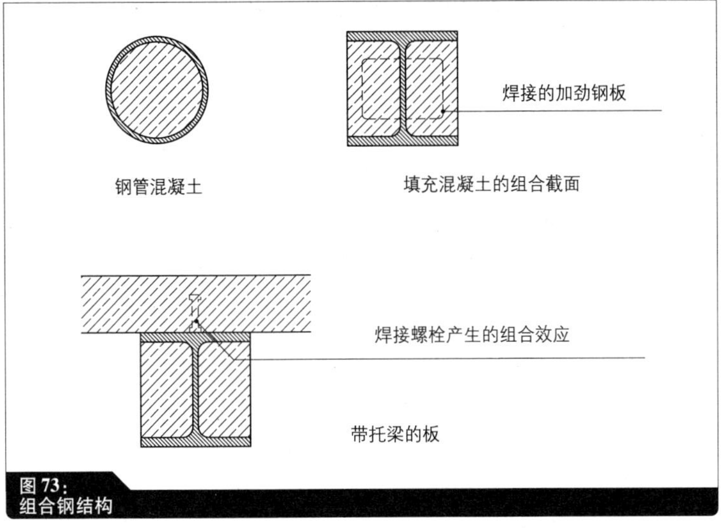

图 73：
组合钢结构

提示：

关于木结构的更多内容参见本套基础教材中的《木结构施工》，中国建筑工业出版社 2010 年出版。

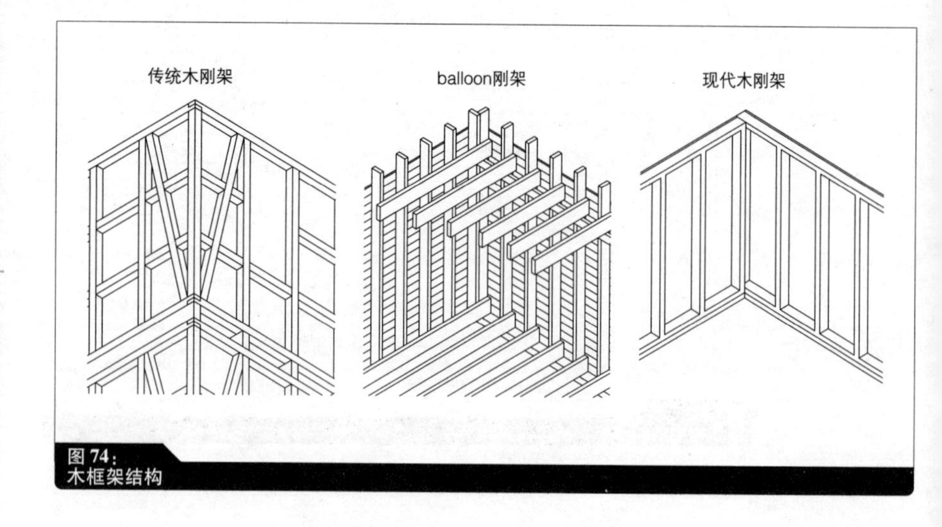

图 74：
木框架结构

Balloon 刚架和平面刚架结构

美国的木结构采用 Balloon 刚架和平面刚架结构。与传统木刚架结构的区别在于它利用了板状木料，这些木料本身无承载力，如果没有墙面的包层将它们固定在其位置上，它们将发生失稳。墙面的包层实际上是起到肋的作用，并与墙体结合在一起形成稳定结构。这种结构有时被称作肋结构。钉子是其主要的连接件。这种结构既经济又灵活。

工程木框架结构

现代工程的木框架结构从静力学讲是一套理想的承重系统，能根据用途的不同而富于变化地建造。薄板和其他板材也正在被使用。

现代的木刚架结构

在木框架结构中，预制件正越来越多地被使用。这里，墙和板构件可以把运输尺寸作为构件的大小。木刚架结构看起来最适合这点。由木板组成且其截面可以由螺栓固定的预制件被制造出来。这些部件也可以以内隔板、面板、窗或门的形式提供，然后组装在一起。与美式结构类似，刚架与薄板板面协同工作，构成了承重系统。(图 74)

P68

支撑构件

当布置框架结构时，主要目标是以楼板和柱来支撑自身重力和竖向荷载，但同时也要注意水平荷载。最重要的水平荷载是风荷载。它能以任何方向作用于建筑上。因为构件间的连接通常为铰接，框架结构几乎无法抵抗水平荷载。因此需要有效的支撑构件。也就是，能将

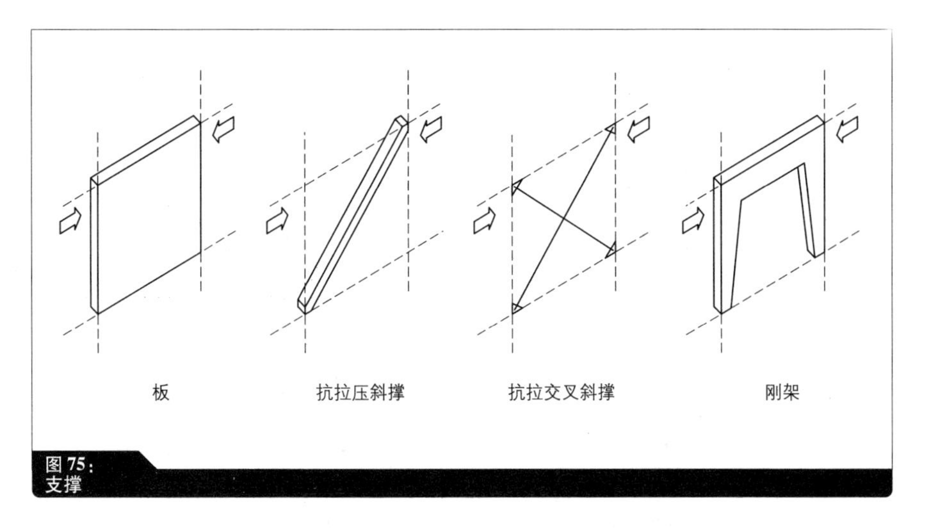

图 75：
支撑

水平荷载从建筑外墙传到基础的一种结构。

支撑构件可以以板的形式发挥作用。它们可以在自身纵向承担水平力，将其向下传导。在高层建筑中，它们作为竖向承载构件将风荷载从各层楼板传递到基础。

板的作用

板可以是实的，用砖石或混凝土制成。这种板的作用也可利用对角斜撑来代替。斜撑在一个方向抵抗压缩，而在另一个方向承受拉伸作用。两根抗拉交叉斜撑也可起到同样的作用（图 75）。在“刚架”一章中也说明了刚架系统的支撑作用。

框架结构必须沿纵向或横向加劲。单独沿一个方向加劲是不够的。因为考虑导平面布置，在一点处两块板总是交叉的。绕交叉点的承载结构可能扭转甚至破坏。为避免这种情况出现，我们需要一个新的支撑面，可以按理想位置布置，但一定不能与其他两个支撑面在同一点交叉（图 76a）。

支撑结构可以不同方式布置，但应靠近中心。因为如果不这样，建筑长边绕该支撑的力臂较长，从而产生较大的力，给支撑带来不必要的变形压力。

楼板

如果框架结构受到水平荷载作用，在同一方向的力，都应该传到一面墙上。这需要一个刚性的楼板，如同图 76 假设的那样。板也可以是带托梁的平板。这种类型的板不是薄板，因为托梁可以产生相对

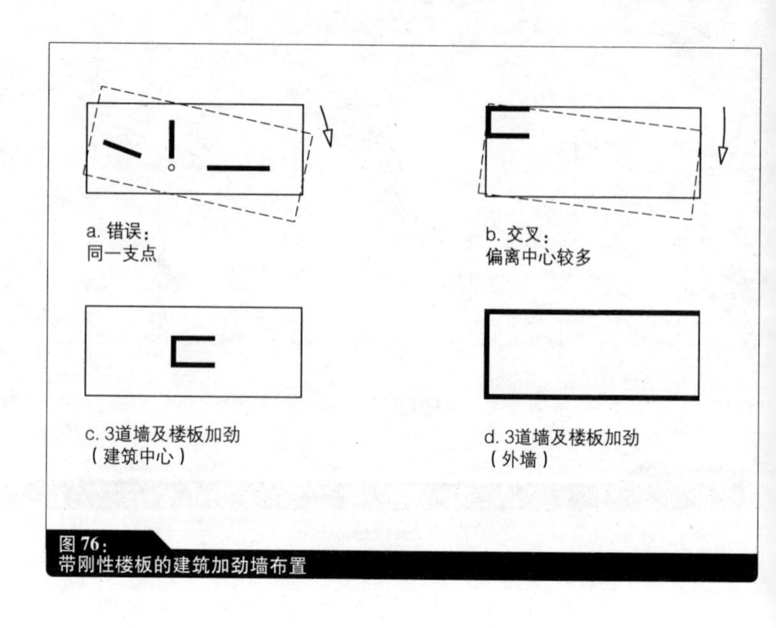

图 76：
带刚性楼板的建筑加劲墙布置

移动。不是所有的水平力都传给支撑结构，中间楼板可以通过加支撑和交叉撑形成刚性板（图 77）。

建筑核心筒

在高层建筑中，带有消防通道和电梯井的建筑核心筒通常用作支撑结构。它们包括最靠近的墙，并且从屋面到基础通长，可以作为竖向承载构件。在高层建筑中，如何抵抗水平力是比承担竖向力更大的

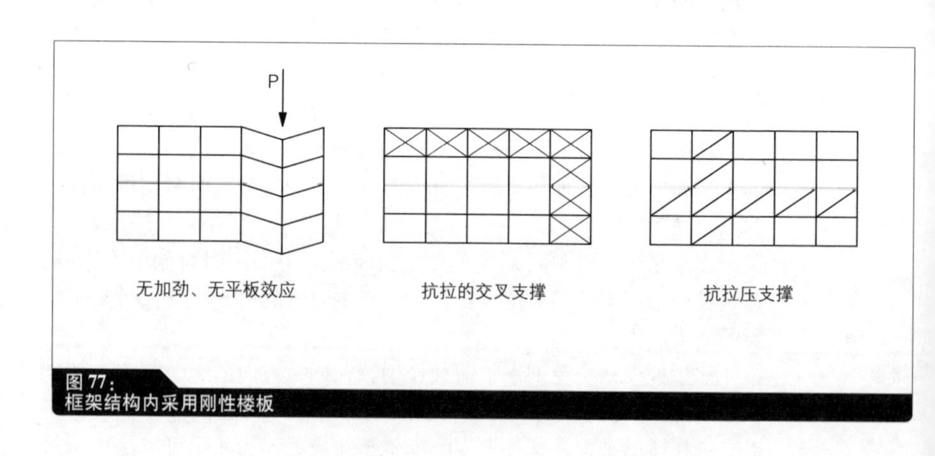

图 77：
框架结构内采用刚性楼板

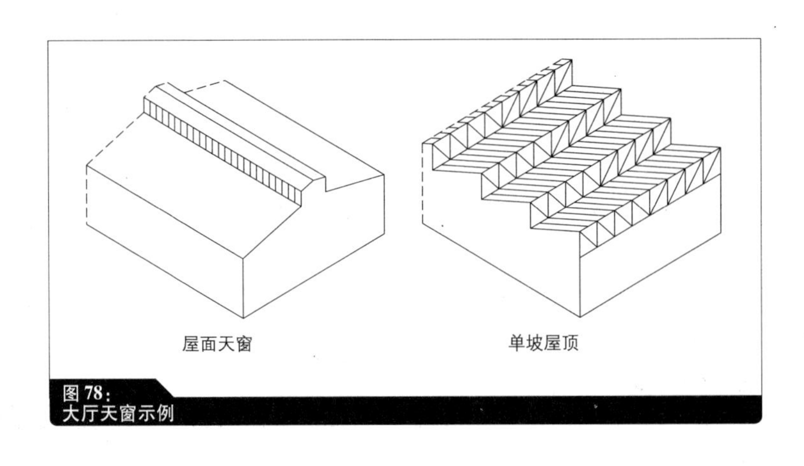

图 78：
大厅天窗示例

问题，因为风速随建筑高度的增加而加大，而风荷载效应也随之加大。在大多数高层建筑中，除建筑核心筒提供支撑外，还有一个可能的方法是将整个建筑外墙形成一个纵向的桁架梁，这样就可以利用最大的梁尺寸，即建筑的整个宽度。

对建筑师来说，主要问题是他们的设计是否可以得到充分的支撑或者换一个方式说，是否稳定。另外，刚度较小和刚度较大的支撑系统之间是有区别的。这取决于支撑如何布置。不同的布置方式会产生不同的效果，这也是刚度不同大小支撑的一个不同。

P70

大厅

大厅的概念是被围护起来的大空间，可以采用任何合理的实体或框架结构形式修建。它们的共同之处在于屋面承重结构跨度较大。屋面的几何尺寸也可以不同的方式设计。重点在于屋面的排水、屋面梁的最优形状或者屋顶天窗的修建方式。

单坡屋顶

天窗可以沿纵向布置，或沿承重结构的方向布置，并作为总体承重结构的一部分，就像单坡屋顶一样。(图 78)

大厅需要大跨度的屋面承重结构。将屋面做成轻质结构是很有利的，因为自重对于整体结构来说是一个外加的荷载。

大厅可以采用不同的结构修建。下面列出了最常用的结构形式。

桁架

屋面刚架或桁架。大跨度梁放置在柱或墙上。由于支座为铰支，这种结构必须以刚屋面，或者外墙，或者柱间，支撑加劲。屋面桁架材料可以为木材、钢材或混凝土。(图 79)

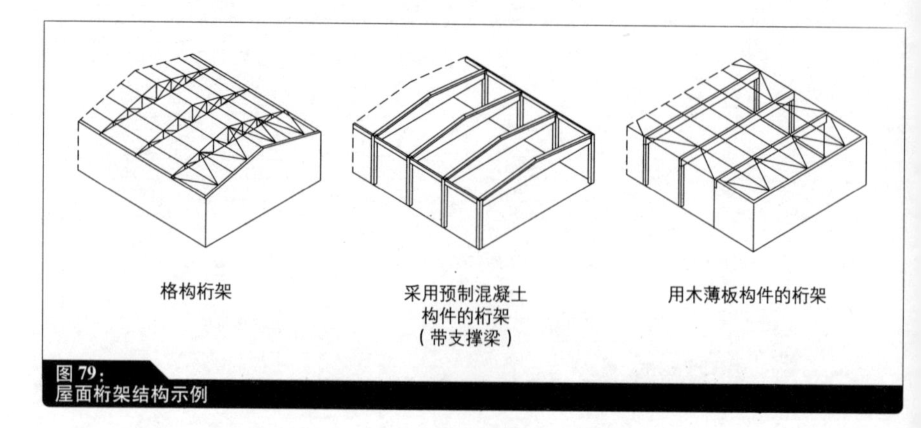

图 79：
屋面桁架结构示例

拱　　因为拱的荷载主要通过轴力而并非弯矩承担，对于大跨来说是一种适合的承重结构，因此也适用于大厅。而对于支座处较大的水平力一定要找到合适的解决办法。拱可以连接在板上，使拱脚推力直接传递到基础，或者放置在柱或墙上，然后布置支撑加强，如墙垛等。也可以在支座间用系杆构件连接，使两边的水平力平衡。这样墙只承担竖向荷载（图 80，“拱”一节）。

刚架　　刚架也很适合大厅结构。它们可用来产生很多几何造型，不像拱那样单调。不对称形式也可以用两铰或三铰刚架很好地实现。但刚架截面尺寸一定要与弯矩图相对应，需通过特定的几何造型和加载方式确定。（图 81，“刚架”一节）

梁网　　前面介绍的系统都是由沿一个方向跨越空间的梁组成。这些都是定向型系统。也可以采用可以承担多个方向荷载的结构。当几个方向跨度差不多时，沿各个方向抵抗荷载是合理的。这时梁相互交叉，形成梁网。这种梁网可以用不同材料制成。它们可以贯穿托梁并与现浇混凝土顶棚相连，形成一个大的连接体系。对于采用钢材和木材的梁网来说，在组装阶段，每个交叉点的抗弯刚性连接都更为复杂。

三维刚架　　桁架梁也可以形成三维承重系统，叫做三维刚架。可以用杆件和节点设计组合而成。三维刚架通常由预制钢构件制成（图 82）。

支撑　　对大厅的加强和支撑要符合上一章讨论过的原则，但其他原则也要重视。例如，只沿一根轴支撑超过一定尺寸的大厅是不够的。因为承重系统内力在传递走之前的传递距离是很长的，所以整体结构的刚度不够。

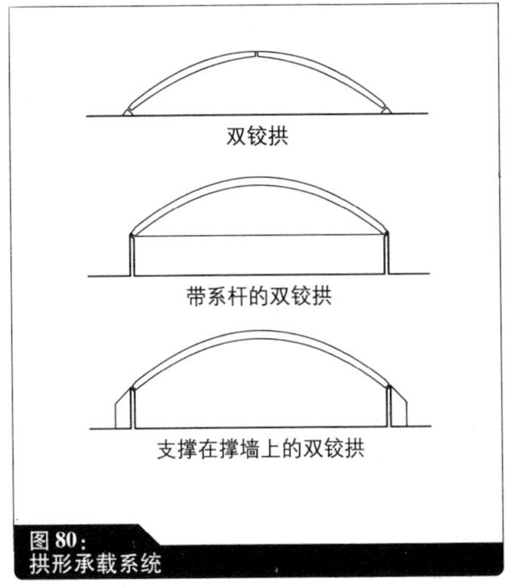

图 80：
拱形承载系统

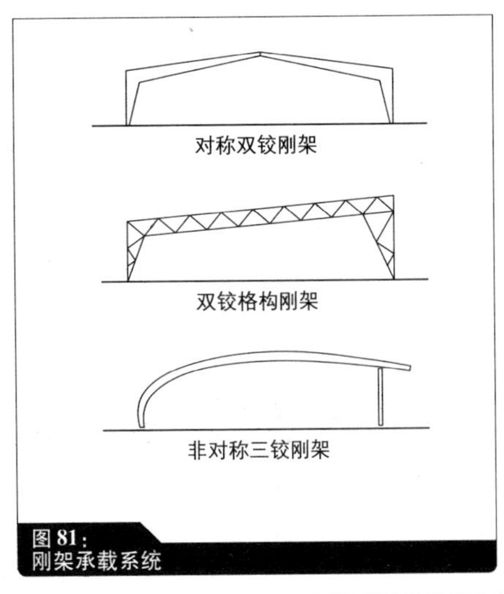

图 81：
刚架承载系统

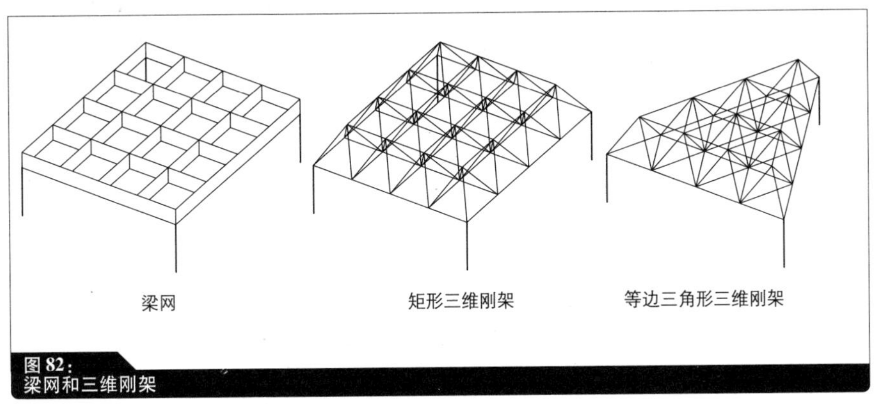

图 82：
梁网和三维刚架

大跨梁容易横向失稳（图 44，115 页），承受很大荷载或山墙承受风载时容易造成结构失效。为防止失稳，通常在屋面水平处，附加一个组合结构支撑山墙，并将荷载传递到屋檐。檩条，沿梁横向方向布置的附加构件，作用于需加强区域，防止其失稳。（图 83，见“支撑构件”）

钢材是大厅结构的理想材料。钢结构轻质、高强，可以用来经济地建造任何静力系统。它的一个优点是可以容易地做出抗弯刚角。

对大厅结构来说，木材也是一种有效的材料。可以利用拱、桁架或用叠层板材制成的刚架，或者实木截面制成的格构桁架。用叠层实

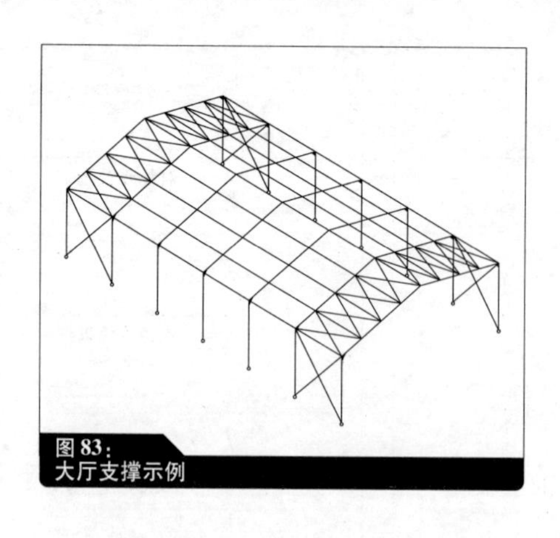
图 83：
大厅支撑示例

木板可以制造抗弯刚性节点。

混凝土大厅通常用预制件修建。它们的承载结构与其他类型不同，柱总是伸入基础内。但桁架是铰接在支座处。钢筋混凝土桁架通常在钢模内生产，尺寸较为固定，所以梁的几何尺寸选择不会太多。

P74

板结构

“结构构件”一节讨论了与梁承受弯矩不同，拱和索是依靠拉压承受荷载的。也可以用板结构三维承载。在设计中，有许多不同的概念和变化因素。下面从总体概念上简要说明一些主要的板结构类型。

褶板/壳

褶板由平面构成并在单独平面内满足“平面效应”，而壳是曲面承重结构，与其他结构有很大不同（图 84）。

像梁一样的板结构

壳或褶板如同梁一样，可以跨越支座间较大的距离。将它们组装在一起就形成了屋顶。在跨越较大距离时，用最小的自重建造最大的高度是非常重要的，曲面或褶板结构非常适合这种情况。它们的承重方式与波纹板非常相似（134 页，图 71）。同时这种板结构必须在边缘处加以支撑，避免由于水平位移引起的倒塌（图 85）。

拉/压承重板结构

与基于拱、索的承重系统不同，板结构有自己特有的承重方式。

穹顶和壳

穹顶、壳以及诸如此类的承重结构在一些区域承受压力，而在另外一些区域承受拉力。周边支撑越连续，它们的承重效果越好。

索网和膜结构

与此相反，所有的悬吊结构，例如索网和膜结构，只是承受拉力。混凝土结构也可以承受拉力。它们可以支撑在周边圈梁或索上。这种类型的索将很大的拉力传递到锚固在基础的支索上。

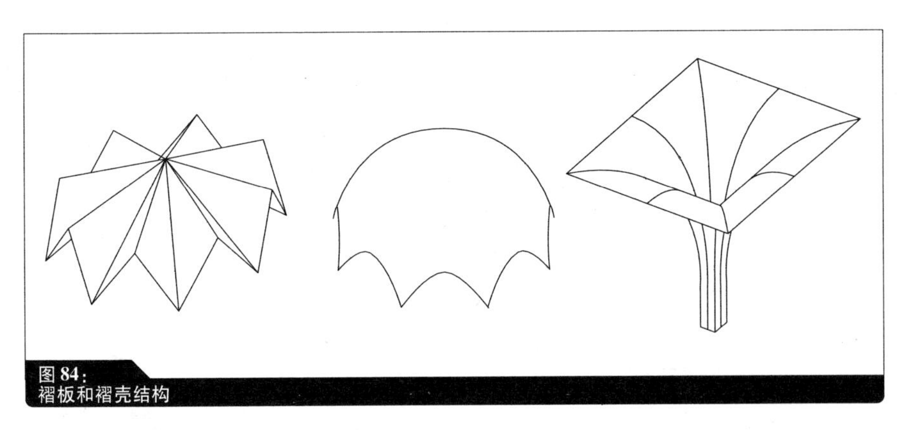

图 84：
褶板和褶壳结构

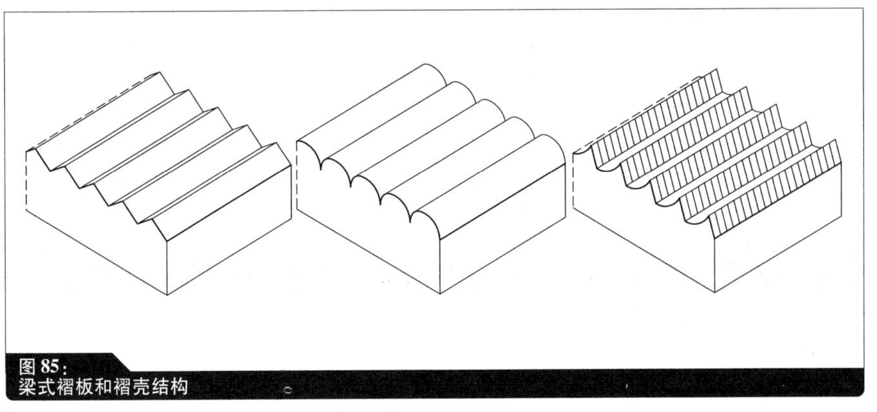

图 85：
梁式褶板和褶壳结构

单曲/双曲面

单曲面是向一个方向弯曲，而在其他方向为线性的曲面。所以可以从平面得到的曲面都是单曲面。它们通常是圆柱或圆锥面的一部分，可以像梁一样在端部支撑或沿长边支撑。与梁不同，沿纵向支撑的单曲面其工作原理与拱一致。

双曲面意味着壳不能由平面形成。图 86 是双曲面的一些示例。双曲面使得结构表面呈现了三维刚性。它保证了受拉平面（如索网和膜结构）如果有足够的预拉力作用，将不会变形。受压圬工或混凝土壳在材料厚度不大的情况下仍然可以形成承载面。

单向或反向弯曲表面

壳或穹顶是沿同一个方向作用的双曲板系统。所有曲线都向一点弯曲。反向弯曲面也叫做鞍形面，通常用于索网和膜结构。

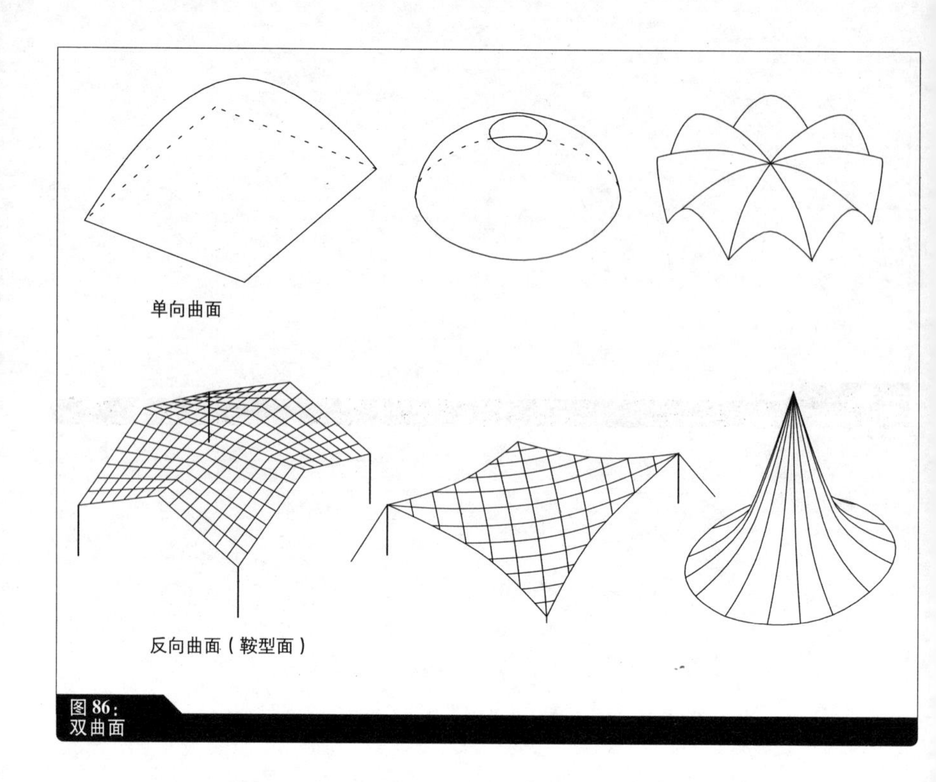

图 86：
双曲面

P77

基础

地基

地基也是承重结构的一部分，如同其他结构构件一样，它必须能够承受所施加的荷载。与其他材料一样，地基土在受载时将产生变形，可能会有几厘米的沉降。沉降是承载行为的正常体现，并不是代表破坏。

与其他工程材料相比，地基土的承载力较低。为了避免超过允许应力，上部结构所传递的荷载必须要分散到足够大的面积上。因此荷载要扩散到较大的地基面中，这意味着基础下的应力随深度增大迅速减小。

地基土类型

有许多类型的地基土，可以不同方式承受荷载。影响地基土性质的主要因素是粒径尺寸和级配。地基土的含水率等湿度指标也是非常重要的。所以收集尽可能多的地基土材料、湿度及地下水位的信息是很重要的。即使是较小的结构在修建时也常需要地基土的研究报告。

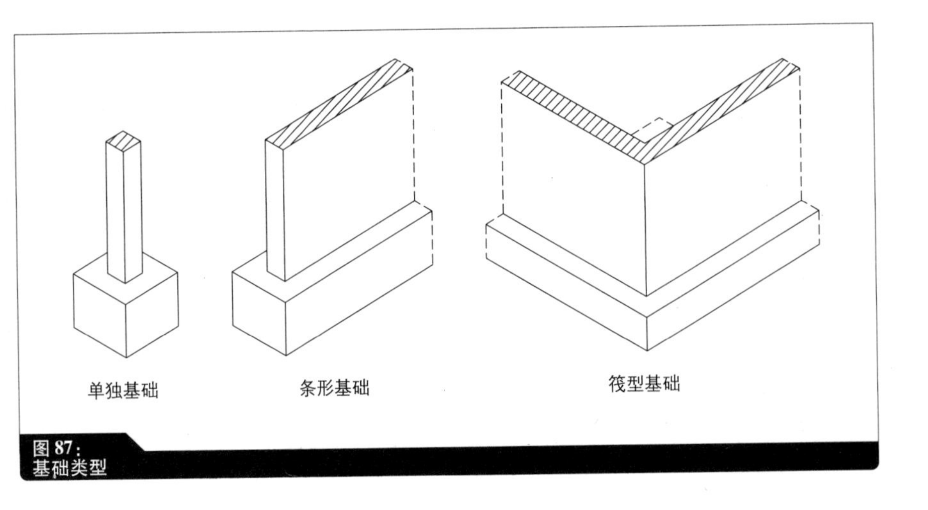

图 87：
基础类型

基础类型

基础起到将荷载传入地基的作用。地基土内的应力取决于荷载分布的面积，即基底尺寸。下面列出了不同基础类型的区别：

—单独基础通常用来承担单根柱的荷载；

—条形基础可以将例如墙的荷载传入地下；

—筏型基础是一片连续的混凝土基底板，可将墙和柱的荷载分布在整个建筑底面积上（图 87）。

基础可以预制，尽管预制对于单独基础来说，花费是比较大的。图 88 显示了一个杯状基础。柱像插入杯口一样插入基础。当将预制柱构件精确地调整到位后，柱和基础间的连接可以用砂浆填充，使二者牢固地连接在一起。

深基础

如果在浅层没有可承载的土层，则必须用深基础传递荷载。可以先钻孔到坚实的土层，然后用混凝土填充。这些钻孔桩如同长柱一样，建筑就修建在上面。荷载主要通过钻孔桩的顶部传递，同时也可以通过粗糙的混凝土桩表面的摩擦来承担荷载。（图 89）

非冻结区域基础

当地基受冻时，由于冰的作用，体积会膨胀。所以在基础下避免冻结是非常重要的。冬天地表下只有一定的深度内会冻结，所以建筑周边的条形连续基础的深度一定要在非冻结区域以内。所要求的深度取决于气候，大概在 80 厘米到 1 米之间。（图 90）

被破坏的基础

沿结构通长出现裂缝通常意味着基础的破坏，破坏的原因可能是建筑结构的不规则或地基的不均匀。

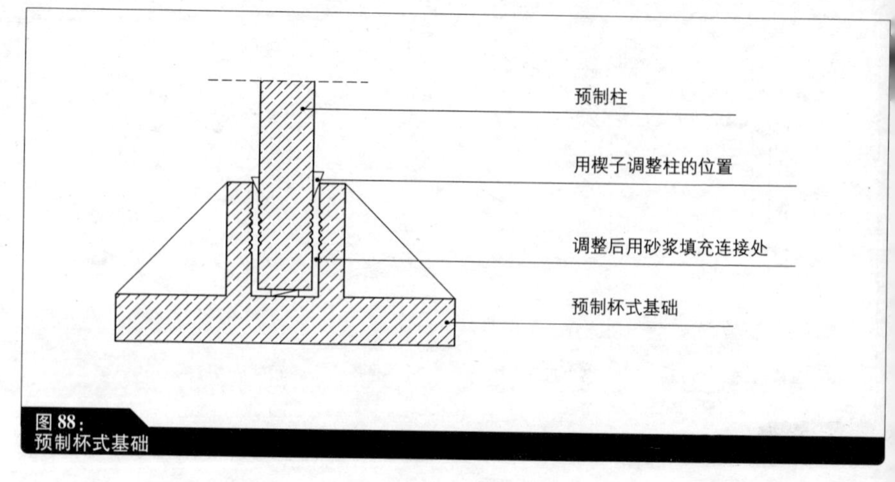

图 88：
预制杯式基础

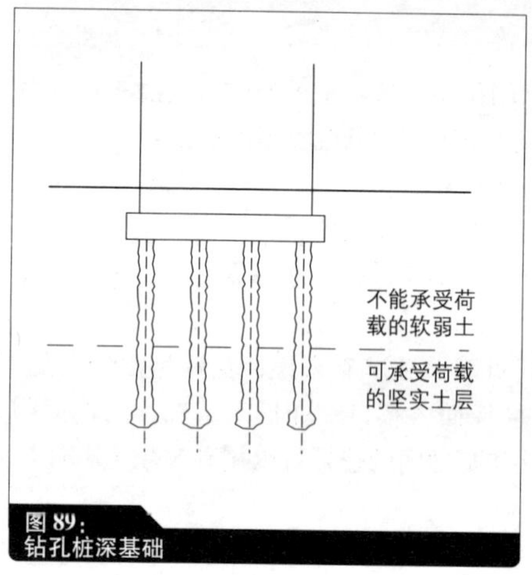

图 89：
钻孔桩深基础

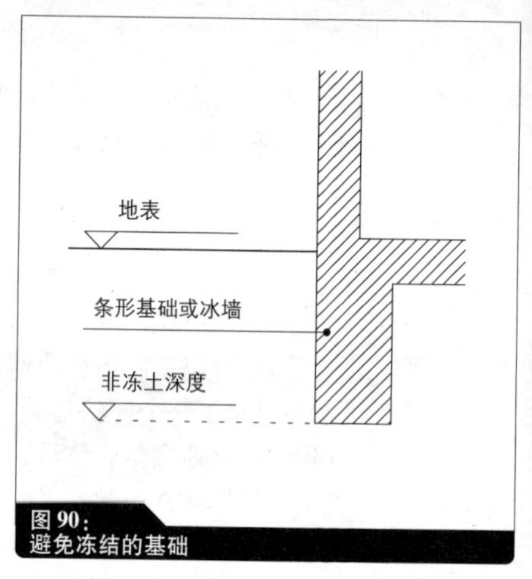

图 90：
避免冻结的基础

地基土性能的变化不可避免地会产生这种问题，因为一般来讲，每种类型的土都有着不同的沉降量。而对结构自身来说，问题产生的原因可能是结构的不同部分对基础施加了不同的荷载或有不同的基础深度，这会导致在基础内产生不均匀的内力。因此应在设计规划阶段就采取措施，或者使荷载均匀地传到土内，或者避免不同沉降量所引起的破坏，比如可以用设置沉降缝的方法来实现。

P81

结语

本书的目的是为学习复杂的承重系统理论开辟一条道路。这里的介绍可使学生了解结构的背景知识，在设计时考虑支撑和荷载的需求，从而能够从实际出发、系统地规划设计。设计承重结构能够帮助设计者塑造具体的空间理念，使他们在承重系统允许的范围内充分发挥自己的创造性。承重系统设计的好坏应从以下方面评估：它是否符合设计者的思路甚至是否有益于设计者的设计思路。这种评估主要是在规划阶段，因为这个阶段的功能和体系使得承重系统成为决定性的因素，例如用到大跨结构时，这种问题只有考虑到承重系统设计的整体复杂性才能够解决。

所以本书介绍的基本知识可以通过对承重系统创造性甚至娱乐性的设计以及根据需求诠释支撑结构的原则，从而拓展成为学生自身建筑设计水平发展的一部分。总结起来，需要考虑三个原则：

1. 承重构件应沿建筑高度通长布置，且在一条直线上。
2. 跨度要尽可能小。大跨度需要大量的人力物力。只有确实需要它们时才能采用。
3. 如果可以允许足够的构件高度的话，那么大跨度结构的处理应该没有问题。即使在这个阶段对结构还不是很了解，但对大跨度应保证有足够的构件高度。

想了解超过本书范畴的承重结构知识的人将会更好地理解结构工程师是如何工作的。同时也能够使学生自己进行计算，提高他们精确确定构件尺寸的能力，使他们能够在承重结构的基础上进行设计。

P82

附录

P82

确定构件尺寸的原则

下面的这些原则可以作为在初步设计阶段确定构件尺寸的一些依据。但它们并不提供承载力方面的数据。

板和顶棚

在多层建筑中平的混凝土板和顶棚：

—最大跨度不超过6.5m；

—该原则应用于简支梁；

—隔声厚度最少16cm；

—点支撑的平板或墙

跨度小于4.3m

$$h\ (\mathrm{m}) \approx \frac{l_i\ (\mathrm{m})}{35} + 0.03\mathrm{m}$$

跨度大于4.3m且板上轻质隔墙允许一定的位移

$$h\ (\mathrm{m}) \approx \frac{l_i^2\ (\mathrm{m})}{150} + 0.03\mathrm{m}$$

木梁板或顶棚

—梁间距70~90cm；

—梁宽≈0.6·d≥cm

梁高 $h \approx \frac{l_i}{17}$

IPE梁

—荷载应绕强轴作用；

—h是截面高，单位cm；q是分布荷载，单位kN/m；l是跨度，单位m。

$$h \approx \sqrt[3]{50 \cdot q \cdot l^2 - 2}$$

HEB梁：

—荷载应绕强轴作用；

—h是截面高，单位cm；q是分布荷载，单位kN/m；l是跨度，单位m。

$h \approx \sqrt[3]{17.5 \cdot q \cdot l^2 - 2}$

大跨屋面支撑结构

分层木板梁（平行）：

—跨度 10～35m；

—桁架间距离 5～7.5m；

高度 $h = \frac{l}{17}$

带平行弦的桁架梁：

—跨度 7.5～60m；

—桁架间距离 4～10m；

总高度 $h \geqslant \frac{l}{12} \sim \frac{l}{15}$

钢实腹梁：

—最大跨度 20m；

—IPE 梁最大高度 600mm。

梁高 $h \approx \frac{l}{30} \sim \frac{l}{20}$

钢桁架梁：

—最大跨度 75m；

梁高 $h \approx \frac{l}{15} \sim \frac{l}{10}$

P84

参考文献

James Ambrose: *Building Structures*, 2nd edition, John Wiley & Sons 1993

James Ambrose, Patrick Tripeny: *Simplified Engineering for Architects and Builders*, John Wiley & Sons 1993

Francis D.K. Ching: *Building Construction illustrated*, 3rd edition, John Wiley & Sons 2004

Andrea Deplazes (ed.): *Constructing Architecture*, Birkhäuser, Basel 2005

Heino Engel: *Structure Systems*, Hatje Cantz, Stuttgart 1997

Thomas Herzog, Michael Volz, Julius Natterer, Wolfgang Winter, Roland Schweizer: *Timber Construction Manual*, Birkhäuser, Basel 2003

Russell C. Hibbeler: *Structural Analysis*, 6th edition, Prentice Hall Publisher 2005

Friedbert Kind-Barkauskas, Bruno Kauhsen, Stefan Polonyi, Jörg Brandt: *Concrete Construction Manual*, Birkhäuser, Basel 2002

Angus J. Macdonald: *Structure and Architecture*, 2nd edition, Architectural Press 2001

Bjørn Normann Sandaker, *The Structural Basis of Architecture*, Whitney Library of Design, New York 1992

G.G. Schierle: *Structure in Architecture*, USC Custom Publishing, Los Angeles 2006

Helmut C. Schulitz, Werner Sobek, Karl-J. Habermann: *Steel Construction Manual*, Birkhäuser, Basel 2000

P85

图片说明

图页 8: Colonnade in front of the Old National Gallery, Berlin, Friedrich August Stüler

图页 34: AEG Turbine Hall, Peter Behrens

图页 58: Berlin Central Station, von Gerkan, Marg und Partner

图 7, 左, 右;
图 41, 左, 中;
图 55, 左, 右: Institut für Tragwerksplanung, Professor Berthold Burkhardt, Technische Universität Braunschweig

其余插图均由作者提供。